SOME PHYSICAL CONSTANTS

Speed of Light	c	3.00×10^8 m/s
Gravitational Constant	G	6.67×10^{-11} N·m^2/kg^2
Coulomb Constant	k	8.99×10^9 N·m^2/C^2
Planck's Constant	h	6.63×10^{-34} J·s
Boltzmann's Constant	k_B	1.38×10^{-23} J/K
Elementary charge	e	1.60×10^{-19} C
Electron Mass	m_e	9.11×10^{-31} kg
Proton Mass	m_p	1.67×10^{-27} kg
Neutron Mass	m_n	1.68×10^{-27} kg

COMMONLY-USED PHYSICAL DATA:

Gravitational Field Strength g	9.80 J/kgm = 9.80 m/s^2
Mass of the Earth M_e	5.98×10^{24} kg
Radius of the Earth R_e	6,380 km (equatorial)
Density of Water	1000.0 kg/m^3 = 1 g/cm^3 *
Density of Air	1.3 kg/m^3 *
Absolute Zero	0 K = −273.15°C = −459.67°F
Freezing point of water **	273.15 K = 0°C = 32°F
Room Temperature	295 K = 22°C = 72°F
Boiling point of water **	373 K = 100°C = 212°F
Normal atmospheric pressure	101.3 kPa

** at normal atmospheric pressure   * at normal pressure, 20°C

USEFUL CONVERSION FACTORS

1 meter = 1 m = 100 cm = 39.4 inches = 3.28 ft
1 mile = 1 mi = 1609 m = 1.609 km = 5280 ft
1 inch = 2.54 cm
1 light-year = 1 ly = 9.46 Pm = 0.946×10^{16} m

1 hour = 1 h = 60 min = 3600 s
1 day = 1 d = 24 h = 86.4 ks = 86,400 s
1 year = 1 y = 365.25 d = 31.6 Ms = 3.16×10^7 s

1 J = 1 kg·m^2/s^2 = 0.239 cal
1 cal = 4.186 J
1 W = 1 J/s
1 kWh = 3.6 MJ

1 K (temperature difference) = 1°C = 1.8°F

$$T \text{ (in K)} = \left(\frac{1\,K}{1°C}\right)[T \text{ (in °C)} + 273.15°C]$$

1.0 radian = 1 rad = 57.3° = 0.1592 revolution

1 m/s = 2.24 mi/h
1 mi/h = 1.61 km/h
1 ft^3 = 0.02832 m^3
1 gallon = 1 gal = 3.79×10^{-3} m^3 ≈ 3.8 kg H$_2$O

1 N = 1 kg·m/s^2 = 1 J/m = 0.225 lb
1 lb = 4.45 N
weight of 1-kg object near earth = 9.8 N = 2.2 lbs

1 cal = energy to raise temp. of 1 g of H$_2$O by 1 K
1 food calorie = 1 Cal = 1 kcal = 1000 cal
1 horsepower = 1 hp = 746 W
1 Pa = 1 N/m^2

$T \text{ (in °C)} = (5°C/9°F)[T \text{ (in °F)} - 32°F]$

$$T \text{ (in K)} = \left(\frac{5\,K}{9°F}\right)[T \text{ (in °F)} + 459.67°F]$$

1 revolution = 360° = 2π radians = 6.28 radians

LATENT HEATS* OF VARIOUS SUBSTANCES† (at standard pressure)

Substance	Melting point (K)	Latent heat* of fusion (kJ/kg)	Boiling point (K)	Latent heat* of Vaporization (kJ/kg)
Helium	(does not solidify at 1 atm)		4.2	21
Hydrogen	14.0	58.6	20.3	452
Oxygen	54.8	13.8	90.2	213
Nitrogen	63.2	25.5	77.4	201
Mercury	234	11.3	630	296
Water	273	333	373	2256
Lead	601	24.7	2013	853
Aluminum	933	105	2720	11400
Copper	1356	205	2840	4730

†Adapted from Halliday, Resnick, and Krane, *Physics,* 4/e, New York: John Wiley, p. 550.

SYMBOLS AND THEIR MEANINGS

$=$	is equal to
$\neq$	is not equal to
$\approx$	is approximately equal to
$>$	is greater than
$<$	is less than
$>>$	is much greater than
$<<$	is much less than
$\equiv$	is defined to be
$\propto$	is proportional to
$\Rightarrow$	implies or therefore
$\Leftrightarrow$	if and only if (implies both ways)
∞	infinity
$\cdot$	indicates a dot product of vectors OR a product of units OR multiplication of scalar quantities
$\times$	indicates a cross product of vectors OR multiplication by a power of 10
i.e.	*id est* "that is"
e.g.	*exempli gratia* "for example"
etc.	*etcetera* "and so on"
Q.E.D.	*quod erat demonstrandum* "which was to be demonstrated"
$|x|$	absolute value of x
mag()	magnitude of a vector
$\int$	indicates an integral
Δ	(as prefix) indicates a largish change in the variable whose symbol follows
Δp	largish momentum transfer
γ	adiabatic index
ε	small unit of energy
λ	wavelength
θ	angle
μ	(as a variable) a quantity that expresses the strength of a magnet
Σ	indicates a sum
ρ	mass density
a	arbitrary constant
A	area (as of a piston)
b	arbitrary constant
c	(often with a subscript) specific heat*
C	constant of proportionality
COP	(not italic) coefficient of performance
°C	(following a number) degrees Celsius
d	(as a prefix) indicates a tiny change in the variable whose symbol follows
$d\!\!\!{}^{-}$	(as a prefix) indicates one contribution to a tiny change in the variable following
$d\!\!\!{}^{-}\vec{p}$	tiny momentum transfer
$d\!\!\!{}^{-}K$	tiny energy transfer
d	distance
e	(as a variable) efficiency of a heat engine
e	(with a superscript) the exponential function
E	total energy
$f()$	some function of whatever is in the ( )
°F	(following a number) degrees Fahrenheit

$\vec{F}$	force
g	multiplicity (almost exclusively in this unit) OR gravitational field strength
h	height OR the Planck constant
$\hbar$	the Planck constant divided by 2π
i	(as a subscript) means *initial* OR represents an index in a sum
J	(not italic) joule, the SI unit of energy
k_s	spring constant
k_B	the Boltzmann constant
K	kinetic energy
K	(not italic) kelvin, the SI unit of temperature
KE	(not italic) kinetic energy
ln	natural logarithm function
L	latent heat*
m	mass
M	mass, usually of a system or a large object
mp	macropartition
n	an arbitrary or unknown integer OR the number of moles in a sample of gas
N	the number of particles in a system or object
N	(not italic) newton, a unit of force
$\vec{p}$	momentum (magnitude is p)
P	pressure (almost exclusively) OR power
Pa	(not italic) pascal, the SI unit of pressure
PE	(not italic) potential energy
Q	heat
$\vec{r}$	position vector
r	radius OR separation OR mag($\vec{r}$)
rms	(as a subscript) root-mean-square
R	the gas constant (in this unit)
S	entropy
t	time
T	temperature
u	an arbitrary variable
U	thermal energy (almost exclusively) OR internal energy in general
$\vec{v}$	velocity
v	speed $\equiv$ mag($\vec{v}$)
V	volume (almost exclusively in this unit) OR potential energy in section T9.6
W	work
x, y, z	position coordinates OR arbitrary variables
x, y, z	(as a subscript) indicates a component of a vector quantity
x-, y-, z-	(as a prefix) indicates a component of a vector quantity

SIX IDEAS THAT SHAPED PHYSICS

Unit T: Some Processes
Are Irreversible

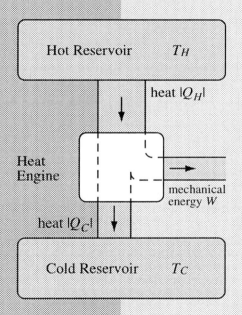

Thomas A. Moore

Pomona College

**WCB
McGraw-Hill**

Boston   Burr Ridge, IL   Dubuque, IA   Madison, WI   New York   San Francisco   St. Louis
Bangkok   Bogotá   Caracas   Lisbon   London   Madrid
Mexico City   Milan   New Delhi   Seoul   Singapore   Sydney   Taipei   Toronto

For Allison,
whose warmth is legendary.

WCB/McGraw-Hill
A Division of the McGraw-Hill Companies

SIX IDEAS THAT SHAPED PHYSICS/
UNIT T: SOME PROCESSES ARE IRREVERSIBLE

This book is printed on recycled, acid-free paper containing 10% postconsumer waste.

2 3 4 5 6 7 8 9 0 QPD/QPD 9 0 9 8

ISBN 0-07-043056-X

Vice president and editorial director: *Kevin T. Kane*
Publisher: *James M. Smith*
Sponsoring editor: *John Paul Lenney*
Developmental editor: *Donata Dettbarn*
Marketing manager: *Lisa L. Gottschalk*
Project managers: *Larry Goldberg, Sheila Frank*
Production supervisor: *Mary E. Haas*
Cover designer: *Jonathan Alpert*
Compositor: *Thomas A. Moore*
Typeface: *9/10 Times Roman*
Printer: *Quebecor Printing Book Group/Dubuque*

Cover photo: © Corel

Library of Congress Catalog Card Number: 97-62247

CONTENTS

PREFACE

1. INTRODUCTION

This volume is one of six that together comprise the PRELIMINARY EDITION of the text materials for *Six Ideas That Shaped Physics,* a fundamentally new approach to the two- or three-semester calculus-based introductory physics course. This course is still very much a work in progress. We are publishing these volumes in preliminary form so that we can broaden the base of institutions using the course and gather the feedback that we need to better polish both the course and its supporting texts for a formal first edition in a few years. Though we have worked very hard to remove as many of the errors and rough edges as possible for this edition, we would greatly appreciate your help in reporting any errors that remain and offering your suggestions for improvement. I will tell you how you can contact us in a section near the end of this preface.

Opening comments about this preliminary edition

Much of this preface discusses features and issues that are common to all six volumes of the *Six Ideas* course. For comments about this specific unit and how it relates to the others, see section 7.

Six Ideas That Shaped Physics was created in response to a call for innovative curricula offered by the Introductory University Physics Project (IUPP), which subsequently supported its early development. IUPP officially tested very early versions of the course at University of Minnesota during 1991/92 and at Amherst and Smith Colleges during 1992/93. In its present form, the course represents the culmination of over eight years of development, testing, and evaluation at Pomona College, Smith College, Amherst College, St. Lawrence University, Beloit College, Hope College, UC-Davis, and other institutions.

The course's roots in the Introductory University Physics Project

We designed this course to be consistent with the three basic principles articulated by the IUPP steering committee in its call for model curricula:

The three basic principles of the IUPP project

1. **The pace of the course should be reduced** so that a broader range of students can achieve an acceptable level of competence and satisfaction.
2. **There should be more 20th-century physics** to better show students what physics is like at the present.
3. **The course should use one or more "story lines"** to help organize the ideas and motivate student interest.

The design of *Six Ideas* was also strongly driven by two other principles:

My additional working principles

4. **The course should seek to embrace the best of what educational research has taught us** about conceptual and structural problems with the standard course.
5. **The course should stake out a middle ground** between the standard introductory course and exciting but radical courses that require substantial investments in infrastructure and/or training. This course should be useful in fairly standard environments and should be relatively easy for teachers to understand and adopt.

In its present form, *Six Ideas* course consists of a set of six textbooks (one for each "idea"), a detailed instructor's guide, and a few computer programs that support the course in crucial places. The texts have a variety of innovative features that are designed to (1) make them more clear and readable, (2) teach you *explicitly* about the processes of constructing models and solving complex problems, (3) confront well-known conceptual problems head-on, and (4) support the instructor in innovative uses of class time. The instructor's manual is much

A summary of the course's distinctive features

more detailed than is normal, offering detailed suggestions (based on many teacher-years of experience with the course at a variety of institutions) about how to structure the course and adapt it to various calendars and constituencies. The instructor's manual also offers a complete description of effective approaches to class time that emphasize active and collaborative learning over lecture (and yet can still be used in fairly large classes), supporting this with day-by-day lesson plans that make this approach much easier to understand and adopt.

In the remainder of this preface, I will look in more detail at the structure and content of the course and briefly explore *why* we have designed the various features of the course the way that we have.

2.   GENERAL PHILOSOPHY OF THE COURSE

Problems with the traditional intro course

The current standard introductory physics course has a number of problems that have been documented in recent years. (1) There is so much material to "cover" in the standard course that students do not have time to develop a deep understanding of any part, and instructors do not have time to use classroom techniques that would help students really learn. (2) Even with all this material, the standard course, focused as it is on *classical* physics, does not show what physics is like *today*, and thus presents a skewed picture of the discipline to the 32 out of 33 students who will never take another physics course. (3) Most importantly, the standard introductory course generally fails to *teach physics*. Studies have shown that even students who earn high grades in a standard introductory physics course often cannot

1. apply basic physical principles to realistic situations,
2. solve realistic problems,
3. perceive or resolve contradictions involving their preconceptions, or
4. organize the ideas of physics hierarchically.

What students in such courses *do* effectively learn is how to solve highly contrived and patterned homework problems (either by searching for analogous examples in the text and then copying them without much understanding, or by doing a random search through the text for a formula that has the right variables.) The high pace of the standard course usually drives students to adopt these kinds of non-thinking behaviors even if they don't want to.

The goal: to help students become competent in using the skills listed above

The goal of *Six Ideas* is to help students achieve a meaningful level of competence in each of the four thinking skills listed above. We have rethought and restructured the course from the ground up so that students are goaded toward (and then rewarded for) behaviors that help them develop these skills. We have designed texts, exams, homework assignments, and activity-based class sessions to reinforce each other in keeping students focused on these goals.

The focus is more on skills than on specific content

While (mostly for practical reasons) the course does span the most important fields of physics, the emphasis is *not* particularly on "covering" material or providing background vocabulary for future study, but more on developing problem-solving, thinking, and modeling skills. Facts and formulas evaporate quickly (particularly for those 32 out of 33 that will take no more physics) but if we can develop students' abilities to think like a physicist in a variety of contexts, we have given them something they can use throughout their lives.

3.   TOPICS EXPLORED IN THE COURSE

The six-unit structure

Six Ideas That Shaped Physics is divided into six units (normally offered three per semester). The purpose of each unit is to explore in depth a single idea that has changed the course of physics during the past three centuries. The list below describes each unit's letter name, its length (1 d = one day ≡ one 50-minute class session), the idea, and the corresponding area of physics.

First Semester (37 class days excluding test days):
Unit C (14 d) *Conservation Laws Constrain Interactions* (conservation laws)
Unit N (14 d) *The Laws of Physics are Universal*   (forces and motion)
Unit R   (9 d) *Physics is Frame-Independent*   (special relativity)

Second Semester (42 class days excluding test days):
Unit E (17 d) *Electromagnetic Fields are Dynamic*   (electrodynamics)
Unit Q (16 d) *Particles Behave Like Waves*   (basic quantum physics)
Unit T   (9 d) *Some Processes are Irreversible*   (statistical physics)

(Note that the spring semester is assumed to be longer than fall semester. This is typically the case at Pomona and many other institutions, but one can adjust the length of the second semester to as few as 35 days by omitting parts of unit Q.)

Dividing the course into such units has a number of advantages. The core idea in each unit provides students with motivation and a sense of direction, and helps keep everyone focused. But the most important reason for this structure is that it makes clear to students that some ideas and principles in physics are more important than others, a theme emphasized throughout the course.

The non-standard order of presentation has evolved in response to our observations in early trials. **[1]** Conservation laws are presented first not only because they really are more fundamental than the particular theories of mechanics considered later but also because we have consistently observed that students understand them better and can use them more flexibly than they can Newton's laws. It makes sense to have students *start* by studying very powerful and broadly applicable laws that they can also understand: this builds their confidence while developing thinking skills needed for understanding newtonian mechanics. This also delays the need for calculus. **[2]** Special relativity, which fits naturally into the first semester's focus on mechanics and conservation laws, also ends that semester with something both contemporary and compelling (student evaluations consistently rate this section very highly). **[3]** We found in previous trials that ending the second semester with the intellectually demanding material in unit Q was not wise: ending the course with Unit T (which is less demanding) and thus more practical during the end-of-year rush.

Comments about the non-standard order

The suggested order also offers a variety of options for adapting the course to other calendars and paces. One can teach these units in three 10-week quarters of two units each: note that the shortest units (R and T) are naturally paired with longest units (E and Q respectively) when the units are divided this way. While the first four units essentially provide a core curriculum that is difficult to change substantially, omitting either Unit Q or Unit T (or both) can create a gentler pace without loss of continuity (since Unit C includes some basic thermal physics, a version of the course omitting unit T still spans much of what is in a standard introductory course). We have also designed unit Q so that several of its major sections can be omitted if necessary.

Options for adapting to a different calendar

Many of these volumes can also stand alone in an appropriate context. Units C and N are tightly interwoven, but with some care and in the appropriate context, these could be used separately. Unit R only requires a basic knowledge of mechanics. In addition to a typical background in mechanics, units E and Q require only a few very basic results from relativity, and Unit T requires only a very basic understanding of energy quantization. Other orders are also possible: while the first four units form a core curriculum that works best in the designed order, units Q and T might be exchanged, placed between volumes of the core sequence, or one or the other can be omitted.

Using the volumes alone or in different orders

Superficially, the course might seem to involve quite a bit *more* material than a standard introductory physics course, since substantial amounts of time are devoted to relativity and quantum physics. However, we have made substantial cuts in the material presented in the all sections of the course compared to a standard course. We made these cuts in two different ways.

The pace was reduced by cutting whole topics...

First, we have omitted entire topics, such as fluid mechanics, most of rotational mechanics, almost everything about sound, many electrical engineering topics, geometric optics, polarization, and so on. These cuts will no doubt be intolerable to some, but *something* has to go, and these topics did not fit as well as others into this particular course framework.

... and by streamlining the presentation of the rest

Our second approach was to simplify and streamline the presentation of topics we *do* discuss. A typical chapter in a standard textbook is crammed with a variety of interesting but tangential issues, applications, and other miscellaneous factons. The core idea of each *Six Ideas* unit provides an excellent filter for reducing the number density of factons: virtually everything that is not *essential* for developing that core idea has been eliminated. This greatly reduces the "conceptual noise" that students encounter, which helps them focus on learning the really important ideas.

Because of the conversational writing style adopted for the text, the total page count of the *Six Ideas* texts is actually similar to a standard text (about 1100 pages), but if you compare typical chapters discussing the same general material, you will find that the *density* of concepts in the *Six Ideas* text is much lower, leading to what I hope will be a more gentle perceived pace.

Choosing an appropriate pace

Even so, this text is *not* a "dumbed-down" version of a standard text. Instead of making the text dumber, I have tried very hard to challenge (and hopefully enable) students to become *smarter*. The design pace of this course (one chapter per day) is pretty challenging considering the sophistication of the material, and really represents a maximum pace for fairly well-prepared students at reasonably selective colleges and universities. However, I believe that the materials *can* be used at a much broader range of institutions and contexts at a lower pace (two chapters per three sessions, say, or one chapter per 75-minute class session). This means either cutting material or taking three semesters instead of two, but it can be done. The instructor's manual discusses how cuts might be made.

Part of the point of arranging the text in a "chapter-per-day" format is to bee clear about how the pace should be *limited*. Course designs that require covering *more* than one chapter per day should be strictly avoided: if there are too few days to cover the chapters at the design pace, than chapters will *have* to be cut.

4. FEATURES OF THE TEXT

The texts are designed to serve as students' primary source of new information

Studies have suggested that lectures are neither the most efficient nor most effective way to present expository material. One of my most important goals was to develop a text that could essentially replace the lecture as the primary *source* of information, freeing up class time for activities that help students *practice* using those ideas. I also wanted to create a text that not only presents the topics but goads students to develop model-building and problem-solving skills, helps them organize ideas hierarchically, encourages them to think qualitatively as well as quantitatively, and supports active learning both inside and outside of class.

A list of some of the texts' important features

In its current form, the text has a variety of features designed to address these needs, (many of which have evolved in response to early trials):

1. **The writing style is expansive and conversational**, making the text more suitable to be the primary way students learn new information.
2. **Each chapter corresponds to one (50-minute) class session**, which helps guide instructors in maintaining an appropriate pace.
3. **Each chapter begins with a unit map and an overview** that helps students see how the chapter fits into the general flow of the unit.
4. **Each chapter ends with a summary** that presents the most important ideas and arguments in a hierarchical outline format.
5. **Each chapter has a glossary** that summarizes technical terms, helping students realize that certain words have special meanings in physics.

6. **The book uses "user-friendly" notation and terminology** to help students keep ideas clear and avoid misleading connotations.

7. **Exercises embedded in the text** (with provided answers) help students actively engage the material as they prepare for class (providing an active alternative to examples).

8. **Wide outside margins** provide students with space for taking notes.

9. Frequent *Physics Skills* and *Math Skills* **sections** explicitly explore and summarize generally-applicable thinking skills.

10. **Problem-solving frameworks** (influenced by work by Alan van Heuvelan) help students learn good problem-solving habits.

11. **Two-minute problems** provide a tested and successful way to actively involve students during class and get feedback on how they are doing.

12. **Homework problems** are generally more qualitative than standard problems, and are organized according to the general thinking skills required.

5. ACTIVE LEARNING IN AND OUT OF CLASS

The *Six Ideas* texts are designed to support active learning both inside and outside the classroom setting. A properly designed course using these texts can provide to students a rich set of active-learning experiences.

The *two-minute exercises* at the end of each chapter make it easy to devote at least part of each class session to active learning. These mostly conceptual questions do not generally require much (if any) calculation, but locating the correct answer does require careful thinking, a solid understanding of the material, and (often) an ability to apply concepts to realistic situations to answer correctly. Many explicitly test for typical student misconceptions, providing an opportunity to expose and correct these well-known stumbling blocks.

Active learning using two-minute exercises

I often begin a class session by asking students to work in groups of two or three to find answers for a list of roughly three two-minute problems from the chapter that was assigned reading for that class session. After students have worked on these problems for some time, I ask them to show me their answers for each question in turn. The students hold up the back of the book facing me and point to the letter that they think is the correct answer. This gives me instant feedback on how well the students are doing, and provides me with both grist for further discussion and a sense where the students need the most help. On the other hand, students cannot see each others' answers easily, making them less likely to fear embarrassment (and I work very hard to be supportive).

Once everyone gets the hang of the process, it is easy to adapt other activities to this format. When I do a demonstration, I often make it more active by posing questions about what will happen, and asking students to respond using the letters. This helps everyone think more deeply about what the demonstration really shows and gets the students more invested in the outcome (and more impressed when the demonstration shows something unexpected).

Active demonstrations

The in-text exercises and homework problems provide opportunities for active learning *outside* of class. The exercises challenge students to test their understanding of the material as they read it, helping them actively process the material and giving them instant feedback. They also provide a way to get students through derivations in a way that actively involves them in the process and yet "hides" the details so that the structure of the derivation is clearer. Finally, such exercises provide an active alternative to traditional examples: instead of simply displaying the example, the exercises encourage students to work through it.

The exercises support active reading

The homework problems at the end of each chapter are organized into four types. *Basic* problems are closest to the type of problems found in standard texts: they are primarily for practicing the application of a single formula or concept in a straightforward manner and/or are closely analogous to examples in the text. *Synthetic* problems generally involve more realistic situations, require

The types of homework problems

students to apply *several* concepts and/or formulas at once, involve creating or applying models, and/ or require more sophisticated reasoning. **Rich-Context** problems are synthetic problems generally cast in a narrative framework where either too much or too little information is given and/or a non-numerical question is posed (that nonetheless requires numerical work to answer). **Advanced** problems usually explore subtle theoretical issues or mathematical derivations beyond the level of the class: they are designed to challenge the very best students and/or remind instructors about how to handle subtle issues.

Collaborative recitation sessions

The rich-context problems are especially designed for collaborative work. Work by Heller and Hollenbaugh has shown that students solving standard problems rarely collaborate even when "working together", but that a well-written rich-context problem by its very open-ended nature calls forth a discussion of physical concepts, requiring students to work together to create useful models. I typically assign one such problem per week that students can work in a "recitation" section where can they work the problem in collaborative groups (instead of being lectured to by a TA).

The goal of the course is that the majority of students should ultimately be able to solve problems at the level of the *synthetic* problems in the book. Many of the rich-context problems are too difficult for individual students to solve easily, and the advanced problems are meant to be beyond the level of the class.

The way that a course is structured can determine its success

In early trials of *Six Ideas*, we learned that whether a course succeeds or fails depends very much the details of how the course is *structured*. This text is designed to more easily support a productive course structure, but careful work on the course design is still essential. For example, a "traditional" approach to assigning and grading homework can lead students to be frustrated (rather than challenged) by the richer-than-average homework problems in this text. Course structures can also either encourage or discourage students from getting the most out of class by preparing ahead of time. Exams can support or undermine the goals of the course. The instructor's manual explores these issues in much more depth and offers detailed guidance (based on our experience) about how design a course that gets the most out of what these books have to offer.

6.   USE OF COMPUTERS

Using computers

The course, unlike some recent reform efforts, is *not* founded to a significant degree on the use of computers. Even so, a *few* computer programs are deployed in a few crucial places to support a particular line of argument in the text, and unit T in particular comes across significantly better when supported by a relatively small amount of computer work.

The most current versions of the computer programs supporting this course can be downloaded from my web-site or we will send them to you on request (see the contact information in section 8 below).

7.   NOTES ABOUT UNIT T

How this unit depends on the other units

This particular unit is a relatively short unit that focuses on developing the concepts of temperature and entropy from a microscopic perspective. This unit can be covered in the course essentially any time after unit N. The idea that the harmonic oscillator has evenly-spaced quantized energy levels draws on chapter $Q8$ of unit Q, the discussion of thermoelectric generators in chapter $T9$ will make more sense after a discussion of the photoelectric effect in chapter $Q4$ and the general background provided by unit E, and there are passing references to the heat generated by current flow that also draw on unit E, but on the whole, the needed concepts could be fairly quickly presented in class if units E and Q were not covered before unit T. There is nothing in this unit that draws on unit R.

How to make cuts if necessary

The unit is a fairly organic whole, so any cuts have to come from the end. Chapter $T9$ on practical heat engines is easily omitted (it is provided mostly for the sake of connecting the unit to reality and for the students' interest), and if one is really not interested in *any* applications of the idea of entropy, chapters T7 and T8 could be omitted as well, though I consider these chapters to be part of the goal of the unit. The remaining chapters are all so intertwined that it is impossible to cut much more.

Experience has shown that the ideas discussed in chapter T5 make *much* more sense to students (and also have a greater impact) when some class and homework time is spent using the computer program *StatMech*. (This is the only unit in the course that really almost *requires* the help of a computer program to function well.) Please download this program from the *Six Ideas* web site: it is freeware, so you may distribute it freely.

Use the computer program *StatMech*!

We have found that this unit does provide a nice note on which to end the course. The concepts are not that difficult, but they are important and really have a profound connection with daily life. Moreover, covering a relatively straightforward unit near the end of the term (when life is usually very stressful) helps students keep up and maintain a positive attitude about the course.

8. HOW TO COMMUNICATE SUGGESTIONS

As I said at the beginning of this preface, this is a preliminary edition that represents a snapshot of work in progress. I would greatly appreciate your helping me make this a better text by telling me about errors and offering suggestions for improvement (words of support will be gratefully accepted too!). I will also try to answer your questions about the text, particularly if you are an instructor trying to use the text in a course.

Please help me make this a better text!

McGraw-Hill has set up an electronic bulletin board devoted to this text. This is the primary place where you can converse with me and/or other users of the text. Please post your comments, suggestions, criticisms, encouragement, error reports, and questions on this bulletin board. I will check it often and respond to whatever is posted there. The URL for this bulletin board is:

The *Six Ideas* bulletin board

        http://mhhe.com/physsci/physical/moore

Before you send in an error or ask a question, please check the error postings and/or FAQ list on my *Six Ideas* web site. The URL for this site is:

The *Six Ideas* web site

        http://pages.pomona.edu/~tmoore/sixideas.html

Visiting this site will also allow you to read the latest information about the *Six Ideas* course and texts on this site, download the latest versions of the supporting computer software, and visit related sites. You can also contact me via e-mail at tmoore@pomona.edu.

Please refer questions about obtaining copies of the texts and/or ancillary materials to your WCB/McGraw-Hill representative or as directed on the *Six Ideas* web-site.

How to get other volumes or ancillary materials

9. APPRECIATION

A project of this magnitude cannot be accomplished alone. I would first like to thank the others who served on the IUPP development team for this project: Edwin Taylor, Dan Schroeder, Randy Knight, John Mallinckrodt, Alma Zook, Bob Hilborn and Don Holcomb. I'd like to thank John Rigden and other members of the IUPP steering committee for their support of the project in its early stages, which came ultimately from an NSF grant and the special efforts of Duncan McBride. Early users of the text, including Bill Titus, Richard Noer, Woods Halley, Paul Ellis, Doreen Weinberger, Nalini Easwar, Brian Watson, Jon Eggert, Catherine Mader, Paul De Young, Alma Zook, and Dave Dobson have offered invaluable feedback and encouragement. I'd also like to thank Alan Macdonald, Roseanne Di Stefano, Ruth Chabay, Bruce Sherwood, and Tony French for ideas, support, and useful suggestions. Thanks also to Robs Muir for helping

Thanks!

with several of the indexes. My editors Jim Smith, Denise Schanck, Jack Shira, Karen Allanson, Lloyd Black, and JP Lenney, as well as Donata Dettbarn, David Dietz, Larry Goldberg, Sheila Frank, Jonathan Alpert, Zanae Roderigo, Mary Haas, Janice Hancock, Lisa Gottschalk, and Debra Drish, have all worked very hard to make this text happen, and I deeply appreciate their efforts. I'd like to thank reviewers Edwin Carlson, David Dobson, Irene Nunes, Miles Dressler, O. Romulo Ochoa, Qichang Su, Brian Watson, and Laurent Hodges for taking the time to do a careful reading of various units and offering valuable suggestions. Thanks to Connie Wilson, Hilda Dinolfo, and special student assistants Michael Wanke, Paul Feng, and Mara Harrell, Jennifer Lauer, Tony Galuhn, Eric Pan, and all the Physics 51 mentors for supporting (in various ways) the development and teaching of this course at Pomona College. Thanks also to my Physics 51 students, and especially Win Yin, Peter Leth, Eddie Abarca, Boyer Naito, Arvin Tseng, Rebecca Washenfelder, Mary Donovan, Austin Ferris, Laura Siegfried, and Miriam Krause, who have offered many suggestions and have together found many hundreds of typos and other errors. Finally, very special thanks to my wife Joyce and to my daughters Brittany and Allison, who contributed with their support and patience during this long and demanding project. Heartfelt thanks to all!

                                        Thomas A. Moore
                                        Claremont, California
                                        November 22, 1997

TEMPERATURE

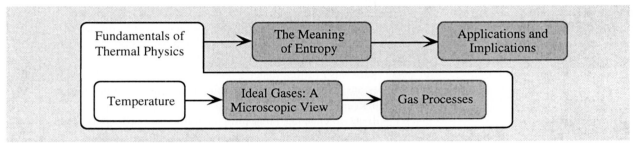

T1.1 OVERVIEW

This unit begins an exploration of one of the greatest mysteries of the universe. Certain physical processes are *irreversible*: a time-reversed version of such a process never seems to occur. For example, thermal energy spontaneously flows from a hot to a cold object, but never the reverse. Living things grow old and die, but never get younger. We remember the past, but not the future. Such processes touch on the mystery of time, indicating its forward direction.

This inexorable movement from past to future becomes even more mysterious when we reflect that the fundamental interactions between subatomic particles in both newtonian and quantum mechanics are *reversible*. How does irreversibility at the macroscopic level arise out of reversible microscopic interactions? In this unit, we will find at least partial answers to these questions.

The first three chapters of this unit lay the foundations for these answers. This chapter sets the stage by talking about the distinction between reversible and irreversible processes and exploring the physical meaning of temperature. Here is an overview of the sections in this chapter.

T1.2 *INTRODUCTION TO THE UNIT* discusses the two approaches that we can take to the physics of complex objects: thermodynamics (a macroscopic approach) and statistical mechanics (a microscopic approach).

T1.3 *IRREVERSIBLE PROCESSES* explores the distinction between reversible and irreversible processes. This section also includes a unit map and a discussion of how the unit is organized.

T1.4 *THE PARADIGMATIC THERMAL PROCESS* describes a simple irreversible process: two objects with initially different temperatures are placed in contact and come into thermal equilibrium. This process raises all of the basic questions that we will explore in this unit and thus will serve for us as a representative of more general irreversible processes.

T1.5 *TEMPERATURE AND EQUILIBRIUM* explores the physically crucial link between temperature and thermal equilibrium.

T1.6 *THERMOMETERS* discusses how we can define a quantitative and physically meaningful temperature scale.

T1.7 *TEMPERATURE AND THERMAL ENERGY* reviews the link between temperature and thermal energy.

T1.2   INTRODUCTION TO THE UNIT

In much of this course to date, we have explored the motion and behavior of objects (from atoms to automobiles to planets) treating them as if they were point-like objects. Yet every object that we experience in our daily life is in fact a complicated system comprised of an enormous number of incredibly tiny particles. We laid some of the foundations for studying complex systems in unit C, but in this unit, we will study such systems in much more depth.

Complex objects have their own distinct physics

Complex systems have physical properties and characteristics that we would not think of assigning to point particles. We have already discussed in Unit C how complex systems have *thermal energy* that must be considered when applying the law of conservation of energy. This thermal energy is related in some (as yet mysterious way) to the object's *temperature*, also something that a point particle does not have. In this unit we will also see that complex systems have a property called *entropy*.

We will see in each case that these quantities are ultimately *statistical* expressions of the internal structure or "state" of a complex system. For example, the link between temperature and the internal state of a system is vividly illustrated by the freezing of water. Clearly something very dramatic happens to the internal state of a system of water molecules when its temperature (whatever that is) is lowered below 0°C.

Physicists used the concepts of *temperature*, *heat*, *thermal energy* and *entropy* to describe the behavior of everyday objects long before they understood such objects to be systems of molecules. The concept of *temperature*, for example, is nearly as old as humanity, since we can directly sense hot and cold. We also know from common experience that when we put a cold object on a fire, something is transferred to it from the fire that makes it hot. Even the idea that this something was a form of energy was discovered in the 1830's, many decades before scientists generally accepted the atomic hypothesis of matter.

Thermodynamics and statistical mechanics

The study of how a macroscopic object's temperature, thermal energy, entropy, and other *macroscopic* characteristics are affected by its interactions with other objects and its general environment is called the study of **thermodynamics**, which became a well-developed science in the middle decades of the 1800s. One of the greatest triumphs of physics in the early decades of the 1900s was the realization that we can explain the thermodynamic behavior of systems in terms of the *statistical* behavior of the molecules as they obey the simple laws of mechanics previously discussed in this course: we call the theory describing how the laws of thermodynamics follow from the basic laws of mechanics **statistical mechanics**. We will focus our attention mostly on the principles of statistical mechanics in this unit.

T1.3 IRREVERSIBLE PROCESSES

Historically, the greatest impediment to this whole approach was a particular stark difference between the behavior of macroscopic objects and the behavior of point particles: the behavior of point particles is generally **reversible**, while the behavior of macroscopic objects is often **irreversible**.

Reversible processes

Imagine, for example, making a movie of two perfectly elastic billiard balls colliding on a pool table. The movie will show the balls approaching each other, exchanging some energy and momentum in the collision process, and then receding from each other. If the movie is run backwards, we would also see the balls approaching each other, exchanging some energy and momentum in the collision process, and then receding from each other. As long as the collision is perfectly elastic (so that the thermal energy of the colliding balls is not involved) there is no way to determine from the events depicted which way the movie should be run.

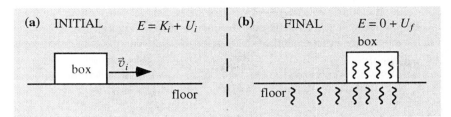

Figure T1.1: The friction interaction between the floor and a sliding box converts kinetic energy to thermal energy. This process is *irreversible*: though the reverse process could be consistent with conservation of energy, it is never observed to occur.

Similarly, imagine a movie of a ball being thrown upward in the air and then caught again as it comes down. As the ball goes up, it converts its initial kinetic energy to potential energy. As it comes down, it does the reverse, converting its potential energy to kinetic energy. If the movie of this process is shown backwards, we again see a ball going up in the air and coming down. Nothing strange or unphysical is observed. If a process looks equally plausible in a movie run backward as in one run forward, it is **reversible**.

This is not accidental: the basic laws of physics (either newtonian or quantum) that describe the interactions of fundamental particles are completely time-symmetric. If such an interaction is physically possible, then a movie of that interaction run in reverse shows an interaction that is also physically possible. There is no forward arrow of time implied by these laws of physics.

The behavior of macroscopic objects can be dramatically different. Consider the simple case of a box sliding across a level floor (see Figure T.1). As the box slides, it slows down due to friction, meaning that its kinetic energy is decreasing. If the floor is level, the box's potential energy is unchanged. Where does its kinetic energy then go? In Unit C, we learned to say that the kinetic energy is converted to thermal energy in the box and the surface on which it slides, and this is reflected by an increase in the temperature of the rubbing surfaces.

Superficially, this looks like a simple energy transformation process similar to the case of the ball being thrown into the air, but consider a movie of this process shown in reverse. A box sitting at rest on the floor spontaneously starts to move, with the rubbing surfaces getting cooler as the box accelerates. Seeing this would make the watchers of the reversed movie really sit up in their seats, because we all know that this simple energy transfer process (internal energy to kinetic energy) does not occur in nature, even though this process is completely consistent with conservation of energy. The conversion of kinetic energy to thermal energy through friction is **irreversible:** the time-reversed process is *not* physically possible.

My professor in an upper-level thermal physics class once handed us a movie projector and told us to make a movie illustrating principles of thermal physics. One group of students (not my group, I regret) made their movie and then rewound it so that it would be shown in reverse. They cleverly intermixed reversible and irreversible processes so that parts of the movie looked completely normal, while other parts were completely outrageous. The movie ended (began?) showing the paper cover of our thermal physics textbook being miraculously created from a heap of ashes in a fire. One of my favorite sequences began with the image of a placid pond on a beautiful spring day. Suddenly, strange ripples began to form on the pond's surface. Growing stronger, these ripples converged to a point on the pond, which suddenly disgorged a large rock. The rock leaped out of the pond into the air, falling into the hands of a surprised student.

The point vividly made by the movie just described is that some processes are just not physically possible in reverse. *Why is this so?* This question becomes even more acute when we seek to explain this irreversible behavior in terms of the interactions between microscopic point-like particles, which (as we've discussed) are *reversible*. This seems logically absurd! How can reversible microscopic processes lead to irreversible macroscopic processes?

Irreversible processes

Ludwig Boltzmann was one of the first physicists to really understand that this contradiction was only apparent. During the 1870's, he published a series of fundamental papers that explained the relationship between the energy, entropy, and the microscopic motions of atoms in interacting macroscopic objects. His work was energetically criticized by many physicists who doubted the reality of atoms and felt that physical theory should not be based on such unobservable and hypothetical entities. Many also dismissed his work out of hand because they could not see how irreversibility could be consistent with reversible microscopic interactions. As a result of struggling against this criticism, Boltzmann became increasingly despondent. He finally committed suicide in 1906, just before experiments with Brownian motion and quantization of electrical charge made it completely clear that the atomic hypothesis was correct.

The great idea of the unit

The "great idea" of this unit is really that Boltzmann was right: *the irreversible behavior of a complex system can be explained by the statistical consideration of the reversible interactions of its molecules.* Our goal in this unit is to develop the ideas and techniques we need to understand this basic idea.

The structure of the unit

A chart illustrating the organization of the unit appears in Figure T1.2. This unit is divided into three subunits. The first subunit (chapters T1 through T3) develops the basic concepts of *temperature, heat*, and *thermal energy* from both a macroscopic and microscopic viewpoint, developing tools and ideas that we will use in the remainder of the unit. In chapters T4 through T6, we will explore the microscopic meaning of *entropy* (which is crucial for describing irreversibility) from the perspective of statistical mechanics. Chapters T7 through T9 discuss some of the applications and implications of the entropy concept, first applying it (in chapter T7) to everyday processes, and then (in chapters T8 and T9) to heat engines, devices explicitly *designed* to convert thermal energy to mechanical energy.

Exercise T1X.1: Which of the following physical processes are reversible and which are (at least approximately) irreversible?
a.   A glass of milk spills on the floor.
b.   An object moves downward, compressing a spring.
c.   Ink poured into a glass of water gradually disperses throughout the water.
d.   A dropped book slams into the floor and remains at rest afterward.
e.   The moon orbits the earth.

Figure T1.2: A chart illustrating the organization of Unit *T*. The boxes outlined in bold at the top are the major subunits; the boxes below each correspond to the chapters in that subunit.

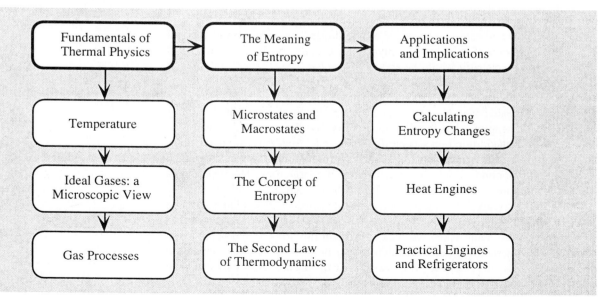

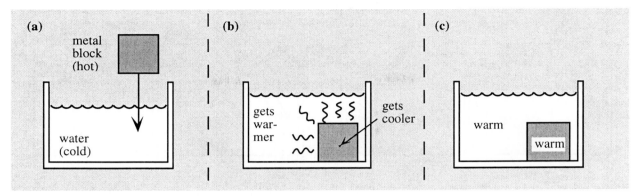

T1.4 THE PARADIGMATIC THERMAL PROCESS

The process of the sliding box vividly illustrates a process that converts energy from one form to another. There is, however, another common irreversible process that will in the long run be more useful to us in explaining the nature of irreversibility.

Imagine dropping a hot piece of metal into a beaker of cold water (see Figure T1.3). We can observe (by touching the metal and water if necessary) that the metal gets cooler and the water gets warmer until their temperatures no longer appear to change. We say that when the temperature of the metal and the water no longer discernibly change the two objects are in thermal equilibrium. Like the sliding box example, this process is irreversible: one never sees a warm metal block placed in warm water spontaneously get hotter while the water becomes colder.

What is happening here? In common language, we might say "heat is being transferred from the metal to the water". It certainly seems like something is flowing out of the metal (leaving it cooler) and into the water (making it warmer). This happens until the metal and water come into thermal equilibrium. In unit *C*, we learned what is being transferred here is in fact *energy*: as energy is transferred from the block to the water, the thermal energy of the block decreases (making it colder) while the thermal energy of the water increases (making it warmer).

This process will be for us the paradigm of an irreversible thermal process. This kind of process is simple and commonplace, and yet it raises all of the big questions that we will have to address in this unit:

1. *What is temperature?* What do we really mean when we say *hot* and *cold*? How is temperature related to internal energy? What properties of a hunk of matter (other than the direct sensation of "hotness" or "coldness") does temperature express?

2. *What is heat?* How is it related to temperature and internal energy? Why does heat spontaneously flow from a hot object to a cold object?

3. *What is thermal equilibrium?* Once energy does begin to flow between the metal and water, why does it stop? What characterizes thermal equilibrium (besides the cessation of changes in temperature)? How is thermal equilibrium related to temperature?

4. *Why is the paradigmatic process irreversible?* Why does heat spontaneously flow from a hot object to a cold object but never the reverse?

As we work through this unit, we will return again and again to this paradigmatic process as we attempt to answer the questions that it raises. The ideas that we develop to explain this particularly simple process will help us to understand more complicated thermal processes as well.

Figure T1.3: The paradigmatic thermal process. **(a)** A hot object (a block of metal here) is placed in contact with a cold object (water here). **(b)** The hot object gets cooler and the cold object warmer **(c)** until their temperatures no longer change: at this point, the objects are in *thermal equilibrium*.

The paradigmatic process

Questions raised by this paradigmatic process

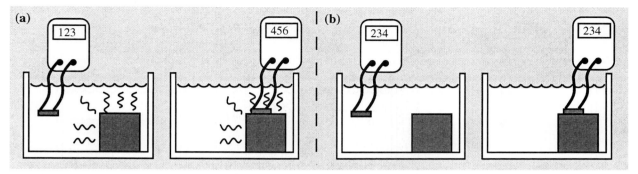

Figure T1.4: Imagine that we place a hot metal block into cold water. **(a)** Before they come into thermal equilibrium they have different temperatures. **(b)** When the block and water come into thermal equilibrium, though, both are always found to have the *same* thermoscope reading.

Quantifying temperature with a thermoscope

The link between temperature and equilibrium

The Zeroth Law of Thermodynamics

T1.5 TEMPERATURE AND EQUILIBRIUM

We can measure temperature qualitatively by touch: we can feel when an object is *hot* and when it is *cold* and can even crudely distinguish degrees of the same. Nonetheless, direct measurement of temperature by touch is subjective and sometimes inconsistent. The first step in making a science out of thermal physics is to figure out how to quantify an object's temperature in an accurate and repeatable manner.

Our crude direct sense of temperature is sufficient for us to discover that the physical characteristics of many objects change as their temperature changes. For example, as temperature increases almost all substances expand, the electrical resistance of a metal changes, the pressure of a confined gas increases, certain kinds of liquid crystals undergo a change of color, and so on.

We could use any one of these temperature-dependent characteristics as a basis for constructing a device that quantifies temperature: we might call such a device a **thermoscope** (we will discuss the difference between a *thermoscope* and a *thermometer* in the next section). For example, we can connect a metal resistor to a digital ohmmeter to construct a simple thermoscope.

When we bring the probe of a thermoscope (the metal resistor in our example thermoscope) in intimate contact with an object at a given temperature, we notice that after a short period of change, the thermoscope reading settles down to a steady value: at this point the probe and object are in thermal equilibrium. The value displayed by the thermoscope thus quantifies the object's temperature.

Experiments then show a curious thing: *two objects are in thermal equilibrium if and only if they have the same temperature* (that is, thermoscope reading). For example, in our paradigmatic thermal process, the temperatures of the water and metal block will change until they have the *same* temperature (that is, until a thermoscope placed successively in contact with each object reads the same number: see Figure T1.4). If the block and water happen to have the same thermoscope reading *before* being placed in contact, then no change results.

The most *apparent* thing about temperature is our sensation of hot and cold. But from the perspective of physics, the most important thing about temperature is that it *characterizes equilibrium*. The so-called **Zeroth Law of Thermodynamics** essentially states that:

A well-defined quantity called **temperature** exists such that two objects will be in thermal equilibrium *if and only if* both have the same temperature.

Surprisingly, this statement seems to be exactly true, independent of the size or physical characteristics of the objects in question. (In chapter T6, we will see how this follows from the principles of statistical mechanics.)

T1.6 THERMOMETERS

A thermoscope as discussed in the last section does not *directly* measure temperature, though. For example, our metal-resistor thermoscope actually registers electrical resistance in a piece of metal, not temperature. Readings registered by a thermoscope with a different piece of metal will be entirely different, even under the same conditions. How can we define a universal temperature scale that everyone can agree on?

We will later see that the Zeroth Law, in conjunction with the concept of entropy, provides us with a means of *mathematically* defining temperature in terms of statistical ideas. But in the absence of this mathematical definition, the best we can do is choose a particular thermoscope to define the temperature scale. A good thermoscope for this purpose will have the following characteristics:

1. It should be based on an easily measured property of a common substance.

2. The chosen property should monotonically increase in value with what our senses and other experiments tell us is "higher" temperature.

3. The property should be measurable over as wide a range of temperatures as possible.

The metal-resistor thermoscope has several flaws in this regard. While it handily satisfies the first item of this list, the resistance of most metals does not monotonically increase with temperature (for a variety of complex reasons). In fact at low temperatures, the resistance of a metal can actually decrease with increasing temperature, reach a minimum, and then increase again. This means that two objects might be measured to have the same thermoscope reading and yet not be in thermal equilibrium. The metal-resistor thermoscope also obviously fails at temperatures above the melting point of the metal.

Characteristics of a good standard thermoscope

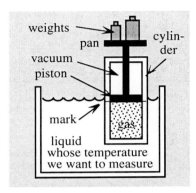

Figure T1.5: An idealized constant-volume gas thermoscope.

Exercise T1X.2: Household thermometers typically measure the expansion of mercury or alcohol in a glass tube. Of the flaws listed above, which do you think is the main flaw of thermometers?

After a certain amount of turmoil and discussion, the physics community eventually decided on the **constant-volume gas thermoscope** as the standard thermoscope that quantitatively defines temperature. A simplified version of such a device is illustrated in Figure T1.5. The temperature registered by such a thermoscope is *defined* to be proportional to the pressure P exerted by the gas when it is in thermal equilibrium with whatever object is being measured:

The constant-volume gas thermoscope

$$T \equiv CP \quad \text{(where } C \text{ is some constant of proportionality)} \qquad \text{(T1.1)}$$

The **pressure** P of a gas (or any fluid) is defined to be the *magnitude* of the force per unit area that it exerts on any surface that separates it from a vacuum. The SI unit of pressure is the **pascal**, where $1 \text{ Pa} \equiv 1 \text{ N/m}^2$. **Standard air pressure** has a defined value of 101.3 kPa: this is approximately equal to the average pressure of the earth's atmosphere at sea level.

In the hypothetical version of the constant-volume gas thermoscope shown in Figure T1.5, a piston confines the gas in a cylindrical container. When we want to measure the temperature of an object, we place the thermoscope in contact with that object and place weights on the piston until the volume of the gas has some specified value (indicated by marks on the cylinder walls). The pressure of the gas when it is confined to this exact volume is the total weight placed on the piston divided by the area of the piston.

As a practical device capable of making precision measurements, this particular design has many flaws, but it nicely illustrates the *principle* of the constant-volume thermoscope. Problem T1R.2 discusses a more practical design.

Having settled on a standard thermoscope, it only remains to set the constant of proportionality in equation T1.1 to determine the temperature scale. We can do this by choosing a value for the temperature of some well-defined and easily reproduced physical situation. Early investigators used the melting and boil-

Defining the Kelvin temperature scale

ing points of water as a standard reference, but it turns out that the exact temperatures of these points is sensitive to various kinds of conditions (including atmospheric pressure) and therefore is imprecisely defined. But it happens that at a certain precise temperature and pressure (0.61 kPa), water can coexist in equilibrium as solid, liquid and gas simultaneously: this is called the **triple-point** of water. This is a well-defined and easily reproducible physical reference point for temperature. By international agreement the triple-point of water is defined to have a temperature of exactly 273.16 K, where the **kelvin** (K) is the SI unit of temperature. The temperature of any object is therefore defined to be:

$$T \equiv 273.16 \text{ K} \left(\frac{P}{P_{TP}} \right) \quad \Rightarrow \quad C \equiv \frac{273.16 \text{ K}}{P_{TP}} \tag{T1.2}$$

where P is the pressure registered by a given gas thermoscope in close contact with the object, and P_{TP} is the pressure registered by that same thermoscope at the triple-point of water.

The temperature of the triple-point was specifically chosen so that a difference in temperature of 1 K almost exactly corresponds to a difference in temperature of 1°C on the old Celsius scale, which was the previous scientific standard. (Note that the kelvin is treated like any other unit in the SI system: specifically, neither the word "degree" nor the degree symbol ° are used for temperatures in K.)

Note also that the constant-volume gas thermometer also defines a temperature of 0 K (where the pressure of the gas is zero). This is an **absolute zero** in the sense that all physically measurable temperatures will be greater than zero (the pressure exerted by a gas can never be negative!). We will learn more about the physical meaning of *absolute zero* in the next chapter.

The zero-density limit

One of the reasons that the constant-volume gas thermoscope was chosen to be the standard is that the reading of such a thermoscope turns out to be fairly independent of the type and amount of gas used. Moreover, the small differences that *do* exist between gases get even smaller as we decrease the density of the gas at the specified constant volume. This is illustrated in Figure T1.6.

To understand this figure, imagine that we have two O_2 (oxygen) gas thermoscopes, one where the density of the gas at the rated volume is 1.0 kg/m^3 and one where it is 0.5 kg/m^3, but which are otherwise identical. If we attempt to measure the temperature of boiling water under standard conditions with these two thermoscopes, we get *different* results (about 373.53 K and 373.33 K respectively)! This kind of ambiguity is *not* desirable. On the other hand, if we plot these readings versus density (black dots on Figure T1.6) and connect them with a straight line, we can extrapolate the reading (373.13 K) that we *would* get if we used O_2 gas of *zero* density. By constructing other O_2 thermoscopes with different operating densities, we can fill in other points on this graph (white dots) and verify that the reading does in fact vary linearly with gas density.

Figure T1.6 shows what we get if we do this not only with our oxygen thermoscope but also with thermoscopes using *other* gases at a variety of densities as well. In each case, the reading for thermoscopes involving a given type of gas seems to depend roughly linearly on the gas density in each case. Even more startling, if we extrapolate the reading for zero density for each type of gas, we get the *same* result (373.13 K) *independent* of the gas used. This strongly suggests that this result has a more fundamental physical significance than the reading given by any particular gas thermoscope operating at a given gas density.

The formal definition of gas-thermoscope temperature is thus

$$T \equiv 273.16 \text{ K} \left(\lim_{\rho \to 0} \frac{P}{P_{TP}} \right) \tag{T1.3}$$

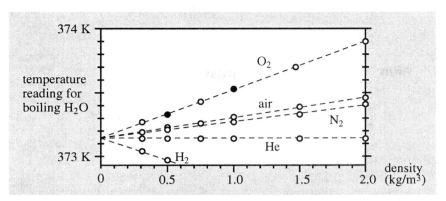

Figure T1.6: The temperature of boiling water (at standard atmospheric pressure) as computed using equation T1.2 for otherwise identical thermoscopes that use different gases with different densities at the specified fixed volumes. (Adapted from Halliday, Resnick, and Krane, *Physics*, 4/e, New York: Wiley, Figure 5, page 499.)

If we use this definition, we do not need to specify the type of gas to use, nor does our gas sample need to be pure (we could even use ordinary air).

Experiments also indicate that the gas temperature so defined seems to be nicely monotonic with physical temperature (so that two objects registered to have different gas temperatures are *never* found to be in thermal equilibrium). Such a thermoscope can also be used to register both extremely high temperatures (up to the melting point of the container) and extremely low temperatures (particularly if the gas is helium) or anywhere in between. An accurate constant-volume gas thermoscope is easily constructed from readily available materials. The fact that all types of gases agree on the reading at the zero-density limit implies that there is something physically significant about that reading that the physical characteristics of the particular gas only slightly modify. There are thus a *number* of practical and theoretical reasons for accepting the constant-volume gas thermoscope as the standard thermoscope.

The formal definition of temperature

While there are a lot of quite practical reasons for settling on this device as the standard thermoscope, I want to emphasize that this choice is nonetheless *conventional*. Another kind of device could (in principle) have been selected to be the standard thermoscope. It is therefore merely fortunate that this practical definition of temperature happens to coincide nicely with the mathematical definition that we will discover later in this unit.

A **thermometer** is any thermoscope that has been calibrated so that when it is placed in contact with an object, it reads a temperature for that object that is *equal* to the extrapolated zero-density reading of a constant-volume gas thermoscope, computed using equation T1.3. The distinction between a *thermoscope* and a *thermometer* is that the latter has been calibrated to be consistent with the accepted standard thermoscope.

A thermometer is a calibrated thermoscope

By design, the size of the kelvin is the same as the Celsius degree, so a temperature *difference* of 1 K = 1°C. The only difference between the two scales is that the zero points are set differently: 0 K corresponds to –273.15°C. So

Converting between temperature scales

$$T_{\text{(in °C)}} = \left(\frac{1°C}{1\,K}\right) T_{\text{(in K)}} - 273.15°C, \qquad T_{\text{(in K)}} = \left(\frac{1\,K}{1°C}\right)[\, T_{\text{(in °C)}} + 273.15°C] \qquad \text{(T1.4)}$$

Because the zero of the Kelvin scale is at the physically meaningful absolute zero, it has a more direct physical meaning than the Celsius scale (as we will see later); this is why we will use the Kelvin scale almost exclusively in this unit. Some "benchmark temperatures" on the Kelvin scale are listed in Table T1.1.

In the United States, we still use the Fahrenheit temperature scale. A temperature *difference* of 5 K = 5°C = 9°F, and absolute zero is –459.7°F. So

$$T_{\text{(in °F)}} = \left(\frac{9°F}{5\,K}\right) T_{\text{(in K)}} - 459.6°F, \qquad T_{\text{(in K)}} = \left(\frac{5\,K}{9°F}\right)[\, T_{\text{(in °F)}} + 459.6°F] \qquad \text{(T1.5)}$$

Center of the sun	1.5×10^7 K	1.5×10^7 °C	2.7×10^7 °F
Surface of the sun	5800 K	5500°C	10,000°F
Melting point of tungsten	3683 K	3410°C	6170°F
Melting point of iron	1808 K	1535°C	2795°F
Melting point of lead	601 K	328°C	622°F
Boiling point of water	373 K	100°C	212°F
Normal body temperature	310 K	37°C	98°F
Room temperature	295 K	22°C	72°F
Freezing point of water	273 K	0°C	32°F
Boiling point of nitrogen	77 K	−196°C	−321°F
Boiling point of helium	4.2 K	−269°C	−452°F
Background temp. of universe	2.7 K	−270.5°C	−454.8°F
Lowest laboratory temperatures	< 0.1 μK	−273.15°C	−459.7°F

Table T1.1: Selected temperature benchmarks.

Exercise T1X.3: On a cold Minnesota winter day, the temperature is −40°F. What is this temperature on the Celsius scale? On the Kelvin scale?

T1.7 TEMPERATURE AND THERMAL ENERGY

Temperature is linked to thermal energy

In unit *C*, we saw that there was a connection between an object's temperature and its internal thermal energy. Empirically, we found that under normal circumstances, if we convert a given amount of kinetic energy to thermal energy in an object, its temperature *increases* by an amount given by the equation

$$dU = mc\, dT \qquad\qquad (T1.6)$$

where *c* is a quantity called the object's **specific heat*** (the asterisk indicating that heat is a misnomer here), *m* is the object's mass, *dU* is the thermal energy added, *dT* is the increase in the object's temperature (where *dU* and *dT* are "sufficiently small" so that *c* is essentially constant). The quantity *c* so defined empirically turns out to be *independent* of the object's size, *strongly* dependent on its composition and phase, and *weakly* dependent on its temperature. Values of *c* range from 4186 J·kg⁻¹·K⁻¹ (for liquid water) to 127 J·kg⁻¹·K⁻¹ (for solid lead). You might review the examples in unit *C* (or look at problem T1S.7) to see how we can use this information to determine equilibrium temperatures.

***U* means "thermal energy"**

By the way, in this unit, *U* will refer *exclusively* to an object's *thermal energy* unless otherwise stated.

Questions raised by the link between *U* and *T*

The idea that thermal energy is indeed *energy* proved very helpful to us in unit *C*, making it possible for us to explain how energy is conserved in situations where it superficially seems to disappear. In our present context, however, this concept raises some perplexing questions:

1. Exactly how is energy stored in an object?
2. Why and how is an object's *temperature* linked to its thermal energy?
3. Why does energy flow spontaneously from high to low temperature (as in our paradigmatic thermal process) but never the other way around?

We superficially discussed the first question in unit *C*, but the others only really arise in the context of this chapter. In this chapter, we have defined temperature in terms of the pressure in a gas thermometer and linked it to equilibrium. What has *either* of these ideas to do with thermal energy?

In the next chapter, we will begin to address the first two questions using a simple microscopic model of a gas: we will see exactly how a gas stores energy and how this energy is related to temperature. The third question is a special case of the general problem of irreversibility, which we will begin to address in chapter T4. In chapter T6 we will finally see how temperature, energy, and entropy are all linked to the concept of equilibrium. Providing full and satisfying answers to these perplexing questions is thus our goal in what follows.

SUMMARY

I. INTRODUCTION
   A. *Thermodynamics* is the study of how an object's temperature, thermal energy, and other macroscopic characteristics are affected by the object's interactions with its environment.
   B. *Statistical mechanics* is a theory that links the laws of thermodynamics to the statistical behavior of molecules.
   C. *Thermal physics* is a general term embracing both of the above.

II. IRREVERSIBLE PROCESSES
   A. Reversible processes
      1. Such processes look possible when seen in a movie run backwards
      2. All fundamental interactions between particles are reversible
   B. Irreversible processes
      1. Such processes are absurd in a movie run backwards
      2. Interactions between macroscopic objects are typically irreversible
   C. Boltzmann claimed that we can explain irreversible macroscopic processes entirely in terms of reversible microscopic processes!
   D. The *paradigmatic thermal process:* Thermal energy spontaneously flows from a hot object to a cold object until they reach equilibrium
      1. This is a simple but crucial example of an *irreversible* process
      2. Studying this process will help us understand Boltzmann's claim

III. TEMPERATURE
   A. Measuring temperature
      1. A thermoscope is any device that quantifies temperature
      2. A thermometer is a thermoscope calibrated to a standard temp. scale
   B. The standard (Kelvin) scale (SI unit of temperature is the *kelvin, K*)
      1. Our standard thermoscope: the *constant volume gas thermoscope*
         a. The temperature T is *defined* to be $\propto$ to the gas pressure p
         b. pressure $p \equiv$ force/area on barrier separating gas from vacuum
      2. The triple-point of H_2O is *defined* to be 273.16 K (sets scale)
      3. So: $T \equiv$ limit as gas density $\to 0$ of (273.16 K)(P/P_{TP}) (T1.3)
      4. this definition turns out to be independent of type of gas used
   C. Temperature scale conversions
      1. The definition of Kelvin scale was chosen so 1 K = 1°C of ΔT
      2. It also sets 0 K to be *absolute zero* (where gas pressure vanishes) instead of at the freezing point of water (0°C = 273.15 K)
      3. so T (in °C) = (1°C/1 K)T (in K) − 273.15°C   (T1.4)
      4. T (in °F) = (9°F/5 K)T (in K) − 459.7°F   (T1.5)
   D. The deepest physical meaning of temperature (very important!):
      1. *Temperature characterizes equilibrium*
      2. The Zeroth Law of Thermodynamics: objects A and B are in thermal equilibrium if and only if $T_A = T_B$.

IV. TEMPERATURE AND THERMAL ENERGY
   A. Temperature T is related to thermal energy U (as discussed in unit C)
   B. Specific heat* c is defined so that $dU = mc\, dT$, where m is mass
      1. c depends strongly on both the type of substance and its phase
      2. c depends weakly on temperature and is independent of mass
   C. These observations leave several open questions, such as
      1. *How* does an object store thermal energy? *Why* is U linked to T?
      2. Why must energy flow from hot to cold and not the reverse?

GLOSSARY

thermodynamics: the study of how temperature, thermal energy, entropy and other macroscopic characteristics of an object are affected by interactions with its surroundings. Thermodynamics describes the general laws governing such interactions without offering any particular explanation for these laws.

statistical mechanics: the theory that explains the laws of thermodynamics in terms of the statistical behavior of the microscopic particles in the objects involved.

thermal physics: the general category that embraces both thermodynamics and statistical mechanics.

reversible process: a process involving an isolated system that remains physically possible if the system's macroscopic initial conditions and final conditions are reversed. A movie of such a process will look reasonable if shown in reverse.

irreversible process: a process involving an isolated system that is physically impossible if the system's macroscopic initial and final conditions are reversed. A movie of such a process will look very strange if reversed.

The paradigmatic thermal process: the process of an initially hot object coming into thermal equilibrium with an initially cold object. This simple process raises most of the fundamental questions about temperature, energy, and irreversibility that we seek to answer in this unit.

thermal equilibrium: the final state of two objects placed in contact. As the objects approach equilibrium, macroscopic properties linked with internal energy and temperature may change with time, but when equilibrium is attained, these properties no longer change: the objects' macroscopic characteristics become stable.

temperature *T*: a physical property of a macroscopic object that can be measured by a thermoscope (see below) and is linked both to the object's thermal energy and thermal equilibrium.

thermoscope: a device that quantifies temperature in terms of changes in some measurable physical property of the working substance in the thermoscope. A bar of metal whose resistance changes with temperature is an example of a thermoscope.

The zeroth law of thermodynamics: the law that asserts that two objects will be in equilibrium *if and only if* they have the same temperature. (Historically, this law was

first stated long *after* the laws known as the first, second, and third Laws of thermodynamics, but as it is *logically* prior to these laws, it was named the *zeroth* law instead of the fourth law of thermodynamics.)

constant-volume gas thermoscope: a thermoscope that quantifies the temperature of its surroundings in terms of the pressure of a quantity of gas held at a fixed volume.

pressure (of a fluid) *P*: the force per unit area exerted by the fluid on a surface separating the fluid from a vacuum.

pascal: the SI unit of pressure. $1 \text{ Pa} \equiv 1 \text{ N/m}^2$.

standard pressure: 101.3 kPa, approximately equal to the average pressure exerted by the earth's atmosphere at sea level.

triple-point of water: a specific temperature and pressure where water can coexist in the solid, liquid, and gas phases simultaneously. This triple-point represents a precisely-defined physical reference point for defining a temperature scale.

the Kelvin temperature scale: the internationally-accepted temperature scale for physics, defined to be proportional to the pressure of a constant-volume gas thermoscope (in the limit of zero gas density). The constant of proportionality is fixed by defining the temperature of the triple-point of water to be 273.16 K. (This makes the freezing point of water at standard pressure 273.15 K).

kelvin: the SI unit of temperature. The definition of the Kelvin scale makes 1 K equal in size to 1°C of temperature difference.

absolute zero: the temperature where the pressure in a constant-volume gas thermometer would be exactly zero. This is the lowest possible temperature.

thermometer: any thermoscope calibrated to read the same temperature that a constant-volume gas thermoscope would under the same conditions.

specific heat* *c*: the quantity *c* in the relationship $dU = mc\,dT$ that characterizes the relationship between changes in an object's temperature and changes in its thermal energy. The asterisk reminds us that "heat" is a misnomer here: we are talking about *thermal energy*. The quantity *c* is independent of the object's mass, depends weakly on the object's temperature, but strongly depends on the substance that the object is made of and the phase of that substance.

TWO-MINUTE PROBLEMS

Note: The answers to *some* of the questions below are debatable! (This is deliberate.)

T1T.1 Characterize each of the following processes as being reversible (A) or irreversible (B).
a.   a living creature grows
b.   a ball is dropped and falls freely downward
c.   a ball rebounds elastically from a wall
d.   a piece of hamburger meat cooks on a grill
e.   a cube of ice melts in a glass
f.   a bowling ball elastically scatters some bowling pins

T1T.2 All irreversible processes involve macroscopic objects (T or F), and always involve transfers to the thermal energy of that object (T or F).

T1T.3 Certain kinds of liquid crystals change color as their temperature changes. A sheet with a layer of such crystals is (A) a thermometer, or (B) a thermoscope. (If the latter, what would you have to do to make the sheet a thermometer?)

T1T.4 Imagine that we place an aluminum cylinder, a wooden block, and a styrofoam cup on a table and leave them there for several hours. We then come back into the room and feel each object. Which (if any) feels coolest? Which (if any) *actually* is coolest?
A. the aluminum cylinder   C. the styrofoam cup
B. the wooden block   D. all are the same

T1T.5 Is pressure a (A) *vector* or (B) *scalar* quantity?

T1T.6 Imagine that we place objects A and B into a large bucket of water and allow them to come into equilibrium with the water. If we now extract A and B from the water and place them in contact with each other, they will *necessarily* already be in equilibrium with each other (T or F).

T1T.7 Your kid brother puts a marble that was in his pocket into his thermos of milk (because it makes a "cool sound" when shaken). If the milk originally had a temperature of exactly 15°C, the temperature of the marble eventually settles down to
A. just a bit below 15°C   D. about 26°C
B. exactly 15°C   E. other (specify)
C. just a bit above 15°C   F. need more information

T1T.8 Imagine that we place a 100-g aluminum block with an initial temperature of 100°C in a styrofoam cup containing 100 g sample of water at 0°C. (The specific heats* of aluminum and water are 900 $J \cdot kg^{-1} \cdot K^{-1}$ and 4186 $J \cdot kg^{-1} \cdot K^{-1}$ respectively). The final temperature of the system will be closest to: A. 0°C   B. 50°C   C. 100°C

HOMEWORK PROBLEMS

BASIC SKILLS

T1B.1 The equation that converts a temperature in °C to a temperature in °F is:

$$\underset{(\text{in } °F)}{T} = \left(\frac{9°F}{5°C}\right) \underset{(\text{in } °C)}{T} + 32°F \qquad (T1.7)$$

Using this equation, show that the melting point of water (0 °C) is 32°F on the Fahrenheit scale, and that the boiling point of water (100°C) corresponds to 212°F. What is the equation that converts a temperature in °F to °C? The hottest recorded daytime temperature on the earth is about 130°F. What is this temperature in Celsius? In kelvins?

T1B.2 The lowest officially recorded temperature within the continental United States is about –70°F. What is this in Celsius? Kelvins?

T1B.3 In Figure T1.5, how much weight would we have to put on the pan if the pressure of the gas is 90 kPa and the area of the piston is 3.0 cm²? (Ignore the piston's weight.)

T1B.4 In Figure T1.5, if you put 1.2 g of weight on the pan and the piston has an area of 2 cm², what is the pressure of the gas? (Ignore the weight of the piston itself.)

T1B.5 Imagine that we measure the pressure of the helium gas in a constant-volume gas thermoscope to be 32.0 kPa at the triple-point of water and 42.3 kPa when immersed in a certain liquid. What is the temperature of that liquid?

T1B.6 Imagine that we measure the pressure of the helium gas in a constant-volume gas thermoscope to be 55.0 kPa at the triple-point of water and 42.3 kPa when immersed in a certain liquid. What is the temperature of that liquid?

SYNTHETIC

T1S.1 Your lab partner claims that physicists have it all backwards: cold thing actually have more thermal energy than hot things, and energy actually flows from cold to hot. What evidence might you point out that would contradict this assertion?

T1S.2 Do we have to represent hotter temperatures by higher numbers, or is this just a convention? If is not a convention, explain why hotter temperatures necessarily must be represented by higher numbers. If this is a convention, suggest why people might have been prompted to choose this convention.

T1S.3 The pressure in a body of water increases with depth. To see exactly *how* the pressure increases, consider the diagram shown in Figure T1.7. Imagine that the water is initially at rest. We then surround a certain block of water with an imaginary rigid plastic cylinder of essentially zero mass, isolating the water inside the cylinder from that outside. Simply isolating the water this way should not cause it to move: if it was at rest before being enclosed it will remain at rest. If it *is* at rest, though, the net force on the cylinder must be zero. Since the water's gravitational interaction with the earth exerts a downward force on the cylinder equal to the weight of the water inside the cylinder, there must be some other force pushing the water upward. This can only be because the water pressure pushing upward on the bottom face of the cylinder is greater than the water pressure pushing downward on the top face. If the area of each cylindrical face is A, then the downward force on the upper face has a magnitude of P_1A (where P_1 is the pressure at depth z_1) and the upward force on the lower face has a magnitude of P_2A (where P_2 is the pressure at depth z_2). Argue that for the water in the cylinder to remain stationary P_2 and P_1 have to be related as follows

$$P_2 - P_1 = \rho g(z_2 - z_1) \qquad (T1.8)$$

where ρ is the density of water. This equation implies that the change in pressure is proportional to the change in depth. (Note that z increases as we go *downward* here.)

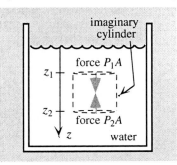

Figure T1.7: The situation for problem T1S.3.

T1S.4 The pressure at the top of a body of water on the surface of the earth is simply the atmospheric pressure P_a there. At what depth below the surface is the water pressure on your body $P = 2P_a$? (Hint: Consider problem T1S.3. Assume that $P_a = 101.3$ kPa and $\rho = 1000$ kg/m³.)

T1S.5 (a) *Argue* using equation T1.8 that if the pressure at the top surface of a fluid is zero, the downward pressure on a horizontal surface of area A at the bottom of the fluid is equal to W/A, where W is the magnitude of the weight of the fluid enclosed by an imaginary vertical column whose base has area A. **(b)** Use this principle to answer the following question: how thick would the earth's atmosphere be if the density of air did not vary with altitude? (Assume that the pressure and density of air at the earth's surface is 101 kPa and about 1.3 kg/m^3 respectively.)

T1S.6 A piston with an area of 1.5 cm^2 slides freely in the middle of a sealed cylinder (see Figure T1.8). The pressure of the gas on the left is 150 kPa, while the pressure of the gas on the right is 120 kPa. What is the magnitude and direction of the force you would have to manually apply to the piston to keep it at rest?

Figure T1.8: The situation for problem T1S.6.

T1S.7 Imagine that we place object A with high initial temperature T_A in contact with object B with low initial temperature T_B. We know from experience that thermal energy will flow from A to B under these circumstances. If these objects are isolated from everything else during this process, then any energy lost by A will be gained by B:

$$dU_A = -dU_B \qquad (T1.9)$$

(a) If this change of thermal energy is "sufficiently small" so that the objects' specific heats* are approximately constant over the temperature ranges involved, show using this equation and equation T1.6 that

$$-\frac{dT_B}{dT_A} = \frac{m_A c_A}{m_B c_B} \qquad (T1.10)$$

(b) Now, since both objects eventually come to the same final temperature T_f, the total change in the temperature of A is $dT_A = T_f - T_A$ and that for B is $dT_B = T_f - T_B$. To save writing, let us define $u \equiv m_A c_A / m_B c_B$. Plug these relations into equation T1.10 and show (after some algebra) that

$$T_f = T_B + \frac{u}{1+u}(T_A - T_B) \qquad (T1.11)$$

(c) Note that if object A is much more massive than object B, then u will be very large and $u/(1+u) \approx 1$, meaning that $T_f \approx T_B + T_A - T_B = T_A$, as one might expect. Argue that equation T1.11 also gives a plausible result if object B is much more massive than object A.

T1S.8 Imagine that we put a 100-g block of aluminum with an initial temperature of 100°C into a cup containing 250 g of water with an initial temperature of 25°C. What is the final equilibrium temperature of this system? (*Hint:* You can use the result of problem T1S.7. The specific heats* of

aluminum and water are 900 J·kg^{-1}·K^{-1} and 4186 J·kg^{-1}·K^{-1} respectively.)

T1S.9 Imagine that we put a 150-g steel ball with an initial temperature of 0°C into a cup containing 150 g of water with an initial temperature of 100°C. What is the final equilibrium temperature of this system? (*Hint:* You can use the result of problem T1S.7. The specific heats* of iron and water are 450 J·kg^{-1}·K^{-1} and 4186 J·kg^{-1}·K^{-1} respectively.)

RICH-CONTEXT

T1R.1 Consider the situation shown in Figure T1.9. Argue that the pressure on the top side of the card times the cross-sectional area of the tube has to be equal to the weight of the water in the tube. The earth's atmosphere exerts pressure on the bottom side of this card. If the latter is to hold the card tight against the bottom of the tube, what is the maximum possible height of water in the tube?

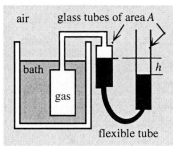

Figure T1.9: The situation for problem T1R.1.

T1R.2 An alternative design for a constant-volume gas thermoscope is shown in Figure T1.10. **(a)** Find an expression for the gas pressure P in terms of the atmospheric pressure P_a, the density of mercury ρ and the height difference h between the mercury in the tubes, and/or the cross-sectional area A of the tubes. **(b)** Describe the advantages of this design over the design shown in Figure T1.5 (list as many advantages as you can).

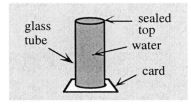

Figure T1.10: To measure the bath's temperature, we can raise or lower the tube on the right until the mercury (black fluid) is exactly at the level of the mark. The gas pressure can then be found from h.

ADVANCED

T1A.1 Imagine that the density of air varies exponentially with altitude according to the following formula

$$\rho(z) = \rho(0)e^{-z/a}$$

Find a value of a that gives the correct air pressure at the surface of the earth. (Hint: study problem T1S.5.)

ANSWERS TO EXERCISES

T1X.1 Reversible: b, e. Irreversible: a, c, d.
T1X.2 The most serious problem with both mercury and alcohol thermometers is their limited range. Mercury freezes at –39°C, which is higher than possible wintertime temperatures in some parts of the United States. Methanol does

not freeze until the temperature falls to –94°C, but boils at 65°C, well below the boiling point of water. Thermometers constructed using these working fluids will thus be useful over only a limited range of temperatures.
T1X.3 –40°F = –40°C = 233 K.

T2

IDEAL GASES

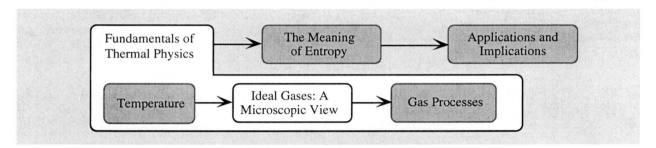

T2.1 OVERVIEW

In the last chapter, we discussed some of the most fundamental concepts in thermal physics. In particular, we saw that:

1. **Temperature expresses equilibrium:** two objects are in thermal equilibrium *if and only if* their temperatures are equal.

2. **Thermal energy is linked to temperature**: an object's internal thermal energy (almost always) increases as its temperature increases.

3. **Energy spontaneously flows from hot to cold**, but is *never* observed to flow spontaneously in the reverse direction!

These concepts raise two important questions: (1) how is thermal energy stored in a complicated system, and (2) why and how is this thermal energy related to temperature? Understanding the answers to these questions is crucial background for the rest of this unit.

We will address these questions in this chapter by exploring the microscopic nature of *gases*, which are the simplest of complex systems. Here is an overview of the sections in this chapter.

T2.2 *THE IDEAL GAS LAW* presents a historical and empirical overview of the *ideal gas law*, which summarizes the observed relationships between various properties of an idealized gas.

T2.3 *A MICROSCOPIC MODEL OF AN IDEAL GAS* develops a microscopic model for a gas that explains the ideal gas law and shows us how a gas' temperature is related to its internal energy.

T2.4 *THE EQUIPARTITION OF ENERGY* discusses the principle of "equipartition" of internal energy and how this helps us understand the specific heats* of real gases.

T2.5 *MOLECULAR SPEEDS AND BROWNIAN MOTION* describes some of the experimental consequences of equipartition and how the molecular model of nature was finally verified.

T2.6 *SOLIDS AND LIQUIDS* presents a qualitative discussion of how energy is stored in solids and liquids, based on what we know about gases.

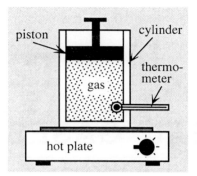

Figure T2.1: Apparatus that we could use to investigate the behavior of gases.

Empirical results of experiments on gases

The ideal gas law

T2.2   THE IDEAL GAS LAW

Consider a gas confined to a cylinder with a piston, as shown in Figure T2.1. By moving the piston up and down, we can vary the volume V in which the gas is confined. Measuring the force that we have to apply to the piston to hold it at rest allows us to determine the pressure P that the gas exerts. The thermometer allows us to measure the gas' temperature T, while the adjustable hot plate allows us to vary that temperature if we want. In short, with such an experimental setup, we could explore the relationships between these variables for any given amount of a gas.

In the late 1650's and early 1660's, Robert Boyle did some of the first known quantitative experiments investigating the well-known fact that gases are much easier to compress than most solids or liquids. In 1662 he reported to the British Royal Society the relationship that is now known as **Boyle's law**: if the temperature of a given volume of gas is constrained to remain constant, the pressure it exerts is inversely proportional to its volume:

$$P = (\text{constant})/V \qquad (\text{fixed } T, \text{ sufficiently low density}) \qquad \text{(T2.1a)}$$

When the volume of the gas is held constant, we have (according to the definition of temperature)

$$P = (\text{constant})T \qquad (\text{fixed } V, \text{ sufficiently low density}) \qquad \text{(T2.1b)}$$

In 1802, Joseph-Louis Gay-Lussac showed that when the *pressure* of a gas is held constant, then:

$$V = (\text{constant})T \qquad (\text{fixed } P, \text{ sufficiently low density}) \qquad \text{(T2.1c)}$$

In 1811, Amedeo Avogadro, building on the work of Gay-Lussac and John Dalton, argued that when pressure and temperature are held constant the volume of a gas is proportional to the number of molecules N in the gas:

$$V = (\text{constant})N \qquad (\text{fixed } P \text{ and } T, \text{ sufficiently low density}) \qquad \text{(T2.1d)}$$

Taken together, these observations imply that (at least at low densities) gases are described by a simple equation that can be written:

$$PV = Nk_BT \qquad (\text{at sufficiently low densities}) \qquad \text{(T2.2)}$$

where k_B is a constant of proportionality called **Boltzmann's constant**, which is determined by empirical measurements to have the *same* value:

$$k_B = 1.38 \times 10^{-23} \text{ J/K} = 8.63 \times 10^{-5} \text{ eV/K} \qquad \text{(T2.3)}$$

for all gases! (*Note*: Boltzmann's constant is often written just k, but considering the many ways that the symbol k is used in physics and mathematics, it seems wise to use the subscript to remind us of which k we are talking about!)

Equation T2.2 is the physicist's version of what is called the **ideal gas law**. Chemists are more used to seeing this equation in the form

$$PV = nRT \qquad \text{(T2.4)}$$

where (using N_A = Avogadro's number = 6.02×10^{23} molecules per mole)

$$n \equiv \frac{N}{N_A} = \text{ number of moles of gas in question} \qquad \text{(T2.5a)}$$

$$R \equiv N_Ak_B = \left(\frac{6.02 \times 10^{23}}{\text{mole}}\right)\left(\frac{1.38 \times 10^{-23} \text{ J}}{\text{K}}\right) = \frac{8.31 \text{ J}}{\text{mole} \cdot \text{K}} \qquad \text{(T2.5b)}$$

(The constant R is often called the **gas constant**.) One of the advantages of *this* form of the law is that n can be determined without actually counting the number of molecules N in a sample of gas

$$n = \frac{M}{M_A} \tag{T2.6}$$

where M is the mass of the sample and M_A is the gas' **molecular mass** (sometimes incorrectly called its *molecular weight*), which is defined to be the mass of one mole (that is, Avogadro's number of molecules) of the gas, typically expressed in units of g/mole. M_A in turn can be calculated by summing the atomic masses (which one can find listed on any Periodic Table) of all atoms appearing in the gas molecule. The value of M_A in g/mole is *approximately* the total number of nucleons (protons and neutrons) in the molecule's atoms. For example, a hydrogen atom (one proton) has an atomic mass of 1 g/mole, an H_2 molecule has a molecular mass of 2 g/mole, a nitrogen atom (7 protons and 7 neutrons) has a molecular mass of roughly 14g/mole, an NH_3 molecule has a molecular mass of 17 g/mole, and so on.

In this chapter, though, we will be focusing on explaining gas behavior in terms of the *microscopic* behaviors of its molecules, so the physicist's version of the ideal gas law (with its explicit reference to the number of molecules N) will be generally more meaningful and helpful. I will consistently use the physicist's version throughout this unit.

Equation T2.2 is called the *ideal* gas law because it is an approximation to the behavior of real gases, an idealization that seems to get better and better as the density of the real gas approaches zero. The ideal gas law thus in some sense expresses the general behavior of gases in the abstract, behavior which is only slightly perturbed by the specific characteristics of a given real gas. The density of gases at atmospheric pressure and room temperature are generally low enough so that deviations from the ideal gas law are measurable but tolerably small.

Exercise T2X.1: Imagine that you have a bottle containing 1400 cm³ of N_2 at atmospheric pressure 101 kPa (1 Pa = 1 N/m²) and room temperature 295 K. How many molecules are in the bottle? What is the mass of the gas?

T2.3 A MICROSCOPIC MODEL OF AN IDEAL GAS

In his *Hydrodynamica* (published in 1738), the Swiss physicist Daniel Bernoulli proposed that the upward pressure exerted by a gas on a piston (as shown in Figure T2.1) might be the effect of multiple impacts of tiny gas particles colliding with the lower surface of the piston. This hypothesis also offered an explanation for the extreme compressibility of gases (compared to liquids or solids), which suggests that there must be a lot of empty space between whatever particles of matter the gas contains. Bernoulli's suggestion was the first step toward a microscopic explanation of the macroscopic behavior of gases.

The purpose of this section is to prove mathematically that the ideal gas law (which for us is so far just an empirical law) emerges very naturally from such a model for gases. In the process of doing this, we will discover something important about the relationship between temperature and internal energy in gases (and it is understanding this relationship that is the ultimate goal of this chapter).

As in the case of models that we have considered previously in this course, the gas model I will present is a simplified picture that makes visualization easier but retains the crucial features of the real system. Since the model is simplified, the results it predicts are likely to be only approximately true, but if the model is good, the approximation will be good. The main purpose of a model, however, is to provide an easy way to *visualize* the system and *reason* about it.

Assumptions of the model

The model of the ideal gas that we will use in this section is based on the following *assumptions*. In each case, I will describe the idealized assumption and then comment on the degree to which it is an approximation to real gases.

1. **A gas consists of a huge number of tiny molecules.** This is a good assumption: most laboratory-sized samples of gas contain on the order of magnitude of 10^{23} molecules, an outrageously large number. Even a bacterium-sized sample of gas contains roughly one molecule for every person in the United States.

2. **Individual molecules are tiny compared to their average separation.** We will essentially treat the molecules as point particles in this model. This assumption is a fairly good description of real gases: the total volume of the molecules in a typical gas is about 1/1000 of the total volume of the gas, implying that most of the gas is empty space. This approximation clearly improves as the density of the gas decreases.

3. **The molecules obey Newton's laws of motion.** This seems superficially not to be a very good assumption: we know from the previous unit that quantum effects are often important for things as small as molecules. It turns out that in this case, however, that since our derivations primarily involve momentum and energy conservation (principles that transcend all theories) the quantum-mechanical treatment of gases yields essentially the same results as a newtonian treatment. The newtonian model, on the other hand, is *much* easier to understand.

4. **The molecules don't interact with each other.** For certain kinds of gases (notably helium and other noble gases) this assumption is very accurate for a wide range of temperatures and densities. In other kinds of gases (notably water vapor) there can be significant interactions between molecules when they are close together: it is these interactions that allow the molecules to form a liquid or solid. The strength of interactions tend to fall off sharply with distance, though, so this assumption accurately applies to any gas whose density is low enough that its molecules spend little time close to their neighbors.

5. **The gas is comprised of identical molecules.** This is just for the sake of simplicity. We will relax this assumption in the section T2.5.

6. **Collisions between the molecules and the container walls are elastic.** Like Assumption 3, this one is not really true, but creating a model where collisions are *not* assumed to be elastic involves much more effort and yet yields the same results. The basic reason is that even though sometimes a molecule does transfer energy to the wall in a collision (or vice versa), when everything is in equilibrium, the *net* energy transferred to or from the wall is zero, so on the average the molecules act *as if* the collisions were elastic. It is easier to just start with this assumption.

7. **The motion of the molecules is entirely random.** If the gas as a whole is at rest in its container (and not sloshing around), this appears to be an excellent assumption. This assumption means that we are just as likely to find a molecule moving in the $+x$ as the $-x$ direction and there is no intrinsic difference between the x, y, and z directions in the gas.

To summarize, assumptions 1 and 7 seem to be well-founded as far as we know, assumptions 3 and 6 are not particularly good but a more rigorous analysis yields the same results, but assumptions 2 and 4 are substantial simplifications that will mean that predictions based on this "ideal gas model" may not accurately reflect real gases under certain circumstances. However, both 2 and 4 become more accurate as the density of the gas decreases.

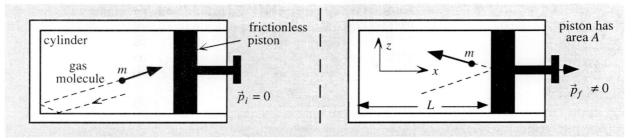

Figure T2.2: When a molecule hits an otherwise isolated piston, it transfers a bit of p-flow to the piston. The total force that the gas exerts on the piston is the rate at which such collisions transfer p-flow to the piston. (In practice, we will exert an external force on the piston to hold it at rest against the force exerted by the gas.)

Now let's see what this model predicts about the behavior of gases. First of all, if the pressure exerted by a gas on its container's walls really is due to gas molecules bouncing off those walls, how can we compute the force exerted by these molecules? We can answer this question with the help of a thought-experiment. Imagine that we confine a sample of gas in a cylinder by a piston that is free to move frictionlessly in response to the gas molecules hitting it, as shown in Figure T2.2. Each collision interaction between a molecule and the piston transfers a certain amount of momentum $d\vec{p}$ to the piston. By definition of force, the total average force exerted by the gas on the piston is the average rate at which these collisions transfer momentum to the piston

$$[\vec{F}]_{avg} \equiv \frac{\Delta \vec{p}}{\Delta t} \qquad (T2.7)$$

(where Δt is long compared to the time between collisions). We can keep the piston from moving in response to this force by exerting on it an opposing force of equal magnitude: in what follows we will assume that we are doing this.

Now, for the sake of simplicity, let us focus on a *single* gas molecule bouncing around the chamber. Each time it hits the piston, we imagine that it rebounds elastically (in accordance with assumption 6), which means that kinetic energy is conserved in the collision. Since we are now holding the piston at rest, conservation of kinetic energy implies that the molecule's *speed* remains the same. What happens in the collision is that the molecule's x-velocity (the component perpendicular to the piston's face) gets reversed (without changing its magnitude $|v_x|$), while the other components are unaffected. The change in the molecule's x-momentum during a collision is thus

$$p_x \text{ (final)} - p_x \text{ (initial)} = -m|v_x| - m|v_x| = -2m|v_x| \qquad (T2.8)$$

The pressure exerted by a single molecule

Conservation of momentum therefore implies that the collision transfers a bit of x-momentum $dp_x = +2m|v_x|$ to the piston.

Now, every time the molecule collides with a wall of the cylinder, either its x-velocity is unaffected (as when the molecule hits an upper or lower wall) or it is reversed without changing magnitude (as when the molecule hits the piston or the opposite wall). In any case, the value of $|v_x|$ is *not* changed by any collision. In between collisions, there are no forces acting on the molecule (assumption 4), so $|v_x|$ remains constant between collisions, too. This means that the time interval between collisions of this molecule with the piston is $2L/|v_x|$, since the molecule must cover a distance of $2L$ in the x-direction (forward and back across the cylinder) between collisions with the piston.

Exercise T2X.2: Argue that this means that the average rate at which the molecule delivers x-momentum to the piston is

$$[F_x]_{avg} = +mv_x^2 / L \qquad (T2.9)$$

Now, the pressure on the piston is defined to be the magnitude of the average force acting on it divided by the piston's area. Dividing both sides of equation T2.9 by the piston's area A, we get

$$P \text{ (for one molecule)} \equiv \frac{[F_x]_{\text{avg}}}{A} = \frac{mv_x^2}{AL} = \frac{mv_x^2}{V} \qquad (T2.10)$$

where $AL = V$ is the volume of the cylinder.

Pressure exerted by N identical molecules

If the molecules don't significantly interact with each other (assumption 4) then the pressure exerted by N molecules will simply be the sum of the pressures due to each individual molecule. Since the molecules do not necessarily have the same speed as they rattle around the cylinder, we cannot simply multiply equation T2.10 by N: we actually have to do the sum. Let P_i be the pressure due to the i-th molecule and let $v_{i,x}$ be the x-velocity of that molecule. Then

$$P \text{ (total)} = \sum_{i=1}^{N} P_i = \sum_{i=1}^{N} \frac{mv_{i,x}^2}{V} = \frac{Nm}{V}\left[\frac{1}{N}\sum_{i=1}^{N} v_{i,x}^2\right] = \frac{Nm}{V}[v_x^2]_{\text{avg}} \qquad (T2.11)$$

Note that in pulling the mass m out of the sum, I have assumed that the molecules are all identical (this is assumption 5), even though I am *not* assuming that all the molecules have the same x-velocities.

If we multiply both sides of this equation by V, we get

$$PV = Nm[v_x^2]_{\text{avg}} \qquad (T2.12)$$

This is very similar in form to the ideal gas law $PV = Nk_BT$. In fact, equation T2.12 *becomes* the ideal gas law if and only if we identify

$$m[v_x^2]_{\text{avg}} = k_BT \qquad \text{or} \qquad \tfrac{1}{2}m[v_x^2]_{\text{avg}} = \tfrac{1}{2}k_BT \qquad (T2.13]$$

(the reason for multiplying by $\tfrac{1}{2}$ will become clear shortly). Now, the model does not *force* us to make this identification. Therefore, to actually *prove* the ideal gas law follows from our model, we would need to add an additional assumption (the empirical fact that the ratio PV/N has the same numerical value for different low-density gases in thermal equilibrium would be sufficient, for example). But we can see that if we simply make this identification, then the ideal gas law emerges from our model in a fairly natural way.

Equation T12.12 actually has some very important implications. Because the gas molecules move in a totally random fashion (assumption 7), the average squared x-velocity of the molecules in the gas should not be different from the average squared y-velocity or squared z-velocity of those molecules:

$$\left[v_x^2\right]_{\text{avg}} = \left[v_y^2\right]_{\text{avg}} = \left[v_z^2\right]_{\text{avg}} \qquad (T2.14)$$

This is simply saying that because the velocities of molecules in the gas are completely random, then no direction in space is different than any other direction. But equations T2.13 and T2.14 together directly imply that:

$$K_{\text{avg}} = \tfrac{1}{2}m[v^2]_{\text{avg}} = \tfrac{1}{2}m[v_x^2 + v_y^2 + v_z^2]_{\text{avg}} = \tfrac{3}{2}k_BT \qquad (T2.15)$$

Exercise T2X.3: Fill in the minor missing steps in equation T2.15.

Equation T2.15 is a very important result in helping us answer the question of how internal energy is related to temperature. Our model implies that the thermal energy of a gas of point particles is really nothing more than the sum of the kinetic energies that the gas' individual molecules have as a result of their random motion. Equation T2.15 asserts that this thermal energy is directly proportional to the temperature T of the gas and depends *only* on N and T (!):

$$U = K_{tot} = N K_{avg} = \tfrac{3}{2} N k_B T \qquad (T2.16)$$

A prediction regarding the thermal energy of a gas

This equation also helps us to understand the physical meaning of *absolute zero*. If temperature is directly proportional to the average kinetic energy of a gas molecule, then absolute zero in a gas implies that the gas molecules all have zero kinetic energy (that is, all the gas molecules are exactly at rest). Clearly, no lower temperature is possible (K_{avg} cannot be negative!).

T2.4 THE EQUIPARTITION OF ENERGY

Equation T2.16 makes a testable prediction about the internal energy of a gas that we might not have suspected without having gone through this argument. The prediction is that if we put a tiny amount of energy dU into a gas, its temperature will increase by dT such that

Reality is more complicated

$$dU = \tfrac{3}{2} N k_B dT \quad \Rightarrow \quad \frac{dU}{dT} = \tfrac{3}{2} N k_B \qquad (T2.17)$$

To test this, we take a sample of gas containing a known number of molecules, put a known amount of energy dU into it, and measure the resulting change in temperature dT. (It turns out that we also have to hold the gas' volume fixed for reasons that we will discuss in chapter T3.) Equation T2.17 states that the ratio dU/dT for one mole of gas should be $\tfrac{3}{2} N_A k_B = \tfrac{3}{2}(6.02 \times 10^{23})(1.38 \times 10^{-23}$ J/K$) = 12.5$ J/K for *all* gases. Do real gases behave this way?

The answer is yes and no. Table T2.1 shows a table of dU/dT for one mole of various kinds of gases. Note that monatomic gases (like helium, argon, and neon) whose molecules are comprised of a single atom do seem to have values reasonably consistent with our prediction. On the other hand, **diatomic** gases (whose molecules consist of two atoms) seem to have a value for dU/dT that is roughly the same across the category but somewhat higher than expected. **Polyatomic** gases (whose molecules are composed of more than two atoms) seem to have a broader range of values that are somewhat higher still.

	dU/dT in J/K
Monatomic	
Helium	12.5
Argon	12.5
Neon	12.7
Krypton	12.3
Diatomic	
Hydrogen (H_2)	20.4
Nitrogen (N_2)	20.8
Oxygen (O_2)	21.1
Carbon monoxide (CO)	21.0
Polyatomic	
Water (H_2O)	27.0
Methane (CH_4)	27.1
Carbon dioxide (CO_2)	28.5

Table T2.1: The value of dU/dT for one mole of various kinds of gases (adapted from R. Serway, *Physics*, 3/e, p. 566.)

Exercise T2X.4: Show that the quoted values of dU/dT for diatomic gases is very nearly equal to $\tfrac{5}{2} N_A k_B$, while the quoted values for polyatomic gases are somewhat higher than $\tfrac{6}{2} N_A k_B$.

Our model *assumes* that the molecules of the gas are pointlike particles. For monatomic gases, this appears to be a good assumption. What the empirical results seem to indicate is that diatomic and polyatomic gases can store thermal energy in forms *other* than the translational kinetic energy of the molecules. For example, a diatomic molecule can in principle rotate and vibrate (something that a point-like particle cannot do) and thus can store energy in the form of rotational kinetic energy, kinetic energy of vibration, and potential energy of vibration. Polyatomic atoms have even more options for energy storage.

We can develop some understanding of what is going on here with the help of two additional models that expand and qualify the basic point particle model. The *equipartition theorem* is a basically newtonian model that lets us predict (at least roughly) how much energy is stored by a molecule. A *quantum limits model* will help us see why some newtonian modes of energy storage may not actually be available to a molecule at a given temperature.

If we write an expression for the total energy of a complicated molecule, (ignoring vibrations) the expression will look something like:

An expression for the total energy of a molecule

$$E_{tot} = \tfrac{1}{2}MV_x^2 + \tfrac{1}{2}MV_y^2 + \tfrac{1}{2}MV_y^2 + \tfrac{1}{2}I_X\omega_X^2 + \tfrac{1}{2}I_Y\omega_Y^2 + \tfrac{1}{2}I_Z\omega_Z^2 \qquad (T2.18)$$

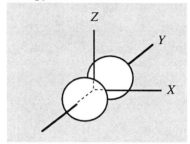

Figure T2.3: An illustration of the three mutually perpendicular axes of symmetry for a diatomic molecule.

where M is the mass of the entire molecule, $\vec{V}$ is the translational velocity of its center of mass, ω_X, ω_Y, and ω_Z are components of its angular velocity of rotation around its three mutually perpendicular axes of effective symmetry X, Y, and Z (see Figure T2.3), and I_X, I_Y, and I_Z are moments of inertia for rotations around those axes. If we consider vibrations as well, we have to add, for each distinct mode of molecular vibration, a pair of terms like $\tfrac{1}{2}mv^2 + \tfrac{1}{2}k_s x^2$ that express the total vibrational kinetic energy of the atoms involved and their total potential energy as a function of displacement from equilibrium in that mode.

I am *not* interested that you understand *anything* about the details of how we would actually write a correct expression for the energy of any given molecule (this is a topic for a higher-level course). What I want you to notice is that, independent of these details, the expression for the total energy of a particular molecule generally involves the sum of terms that all have the same *form*: a constant times the *square* of either a velocity or displacement component. Physicists think of each such quadratic term in the expression for a molecule's total energy as corresponding to a **degree of freedom** for that molecule.

The equipartition theorem

The **equipartition theorem** simply asserts the following:

> On the average, each of a molecule's degrees of freedom stores $\tfrac{1}{2}k_BT$ of energy (unless quantum effects squash it, as we will discuss below).

This theorem generalizes the idea expressed in equations T2.13 and T2.14, where we saw that each of the three degrees of freedom associated with the translational kinetic energy of a molecule stores $\tfrac{1}{2}k_BT$ of energy on the average: it simply asserts that each quadratic term in the energy expression behaves similarly.

Why the equipartition theorem makes sense

This theorem also makes a certain amount of basic sense. Consider a diatomic molecule bouncing around an enclosure. Imagine for the sake of argument that the molecule has zero rotational kinetic energy initially. If the molecule hits a wall a glancing or off-center blow, some of the kinetic energy associated with the translational motion of its center of mass is going to be converted to rotational energy in the collision. On the other hand, a molecule that is rotating very rapidly might transfer some of its rotational kinetic energy back to translational kinetic energy as it hits the wall. On the average, a balance will be achieved where a molecule stores some of its energy in rotational form and some in translational form. Moreover, the energy stored in these forms should be roughly proportional: if the translational motion of molecules were to become more violent, one would expect that the rotational motions would also eventually become more violent as a result of more violent collisions with the walls. The equipartition theorem embraces this common-sense logic and makes it quantitative.

A general mathematical proof of the theorem is beyond our means here (though we will essentially prove it for one specific case in chapter T6). Even so, the theorem is so important for thermal physics that it is worth knowing. For our purposes, we can take it as simply part of our model of gases.

The quantum limits model

If newtonian mechanics were true, *every* quadratic term in the expression for a molecule's energy would *always* get an average energy of $\tfrac{1}{2}k_BT$. However, quantum mechanics means that in some circumstances, some terms do not get their normal share of energy. This can happen in either of two ways.

First limit: rotations that leave a molecule unchanged don't count

First of all, quantum mechanics implies (though I can't prove this here) that if a molecule is completely unchanged by a rotation around a certain axis, then the term for that degree of freedom does not actually appear in the energy expression and thus does not store any energy. This idea alone goes a long way toward explaining the observed results in Table T2.1. Monatomic gas molecules are simply isolated atoms that, because they are spherical, are unchanged by rotation around *any* axis. This means that *none* of the three rotation terms in equation T2.18 apply to monatomic molecules, so each such molecule only has the three translational degrees of freedom and thus stores only $\frac{3}{2}k_BT$ of energy per molecule. Therefore the total energy stored by N molecules of a monatomic gas is $U = \frac{3}{2}Nk_BT$. Diatomic molecules, on the other hand, are unchanged only by rotations around their long axis (the Y axis in Figure T2.3), so only one of the three rotational degrees of freedom doesn't count. This means that the molecule has five degrees of freedom (three translational and two rotational), and N such molecules have an energy of $U = \frac{5}{2}Nk_BT$. Polyatomic molecules are rarely completely symmetrical around any axis, and so generally have six degrees of freedom (three translational and three rotational), so $U = \frac{6}{2}Nk_BT$. In summary:

Type of gas	Degrees of freedom	Energy stored in N molecules
Monatomic	3 translational	$U = \frac{3}{2}Nk_BT$
Diatomic	3 translational + 2 rotational	$U = \frac{5}{2}Nk_BT$
Polyatomic	3 translational + 3 rotational	$U = \frac{6}{2}Nk_BT$

Summary of implications for various types of gases

These predictions match the empirical data pretty well (see exercise T2X.4).

But what about *vibrational* energy terms? A diatomic molecule has one obvious mode of vibration (where the two atoms move toward and away from each other along the long axis, stretching and relaxing the bond between them) and so should have two more degrees of freedom (one vibrational kinetic energy term and one potential energy term), leading to a predicted total energy of $U = \frac{7}{2}k_BT$, instead of $U = \frac{5}{2}k_BT$ as empirically observed. Polyatomic molecules should have even more vibrational degrees of freedom. Why don't these seem to count?

The answer has to do with the second way that quantum mechanics can squash a degree of freedom. The derivation of the equipartition theorem is based on the newtonian *assumption* that collisions can transfer arbitrarily small amounts of energy from any one storage mode to any another. Quantum mechanics, on the other hand, says the energy held by any of a molecule's energy storage modes will be *quantized*. Therefore, energy can only be transferred to or from that storage mode in chunks of a certain specific size.

Second limit: energy quantization can "freeze out" degrees of freedom

If the chunks of energy that have to be transferred to change the energy associated with a certain degree of freedom are too large, that degree of freedom may essentially be "frozen out" from participating in the free interchange of energy assumed by the equipartition theorem, because no collision that the molecule might reasonably participate in has enough energy to boot that storage mode out of its ground state. Consider a hydrogen molecule (a diatomic molecule), for example. At room temperature $k_BT \approx 0.026$ eV $\approx 1/40$ eV, so an average energy available to a hydrogen molecule colliding with a wall is $\frac{3}{2}k_BT \approx 0.039$ eV. The energy levels associated with a molecule's translational motion are typically so close together compared to this that this energy storage mode behaves entirely classically. The lowest rotational energy levels for a hydrogen molecule happen to be about 0.015 eV apart, which is enough smaller than 0.039 eV so that energy can fairly freely be exchanged between translational and rotational modes during a collision. On the other hand, the difference between *vibrational* energy levels for this molecule is about 0.27 eV. This is so large compared to the typical energy available during a collision that even if *all* of the energy in a typical collision were available for conversion to rotational energy, it would not be a big enough chunk to get the vibrational mode out of its ground state. Therefore, the vibrational mode for a hydrogen molecule simply doesn't participate in free interchange of energy (and so for the equipartition theorem, doesn't really exist as a degree of freedom) until the temperature becomes high enough so that k_BT becomes comparable to 0.27 eV (about 3000 K).

As a general rule of thumb, the energy storage mode associated with a degree of freedom will be completely frozen out unless k_BT is roughly equal to half the difference between its lowest energy levels, and is not really fully engaged until k_BT is equal to at least twice that difference (between these limits, the mode stores an energy that is a fraction of $\frac{1}{2}k_BT$, a fraction that grows as T increases).

Rules of thumb for determining if a mode is frozen

In practice, this means that the vibrational degrees of freedom of most gas molecules are frozen out at room temperature (though vibrational modes of massive molecules such as Cl_2 and CO_2 may be partially unfrozen), while translational and rotational degrees of freedom are typically fully engaged at room temperature.

Exercise T2X.5: Verify that $k_B T \approx 0.026$ eV at room temperature = 295 K.

Exercise T2X.6: At roughly what temperature will the rotational degrees of freedom of a hydrogen molecule be completely frozen out?

T2.5 MOLECULAR SPEEDS AND BROWNIAN MOTION

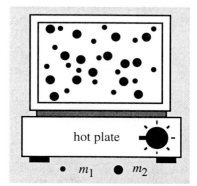

Figure T2.4: A container holding molecules of two different masses m_1 and m_2 sits on a hot plate that we can use to control its temperature.

The rms speed of a molecule in a gas

Brownian motion

The equipartition theorem applies even when we have a mixture of gases. For example, imagine a mixture of two gases, one having molecules with mass m_1 and the other having molecules of mass m_2, as shown in Figure T2.4. The equipartition theorem implies that if the two gases are in thermal equilibrium with the walls and each other, the average translational kinetic energies of the molecules in both gases must be the same:

$$\tfrac{1}{2} m_1 [v_1{}^2]_{\text{avg}} = \tfrac{1}{2} m_1 [v_x^2 + v_y^2 + v_z^2]_{\text{avg}} = \tfrac{3}{2} k_B T = \tfrac{1}{2} m_2 [v_2{}^2]_{\text{avg}} \qquad \text{(T2.19)}$$

This means that the more massive a molecule is, the smaller its average squared speed will be. The small molecules in Figure T2.4 will thus zip around their container rapidly, while the big molecules will amble about at lower speeds.

An approximate measure of the average speed of a molecule is the square root of its average squared speed, which can be easily calculated for a given temperature using $K_{\text{avg}} = \tfrac{1}{2} m [v^2]_{\text{avg}} = \tfrac{3}{2} k_B T$:

$$v_{\text{rms}} \equiv \sqrt{[v^2]_{\text{avg}}} = \sqrt{\frac{3 k_B T}{m}} \qquad \text{(T2.20)}$$

This quantity (which technically is not exactly the same as the *average speed* of the molecule but is a close approximation to it) is often called the molecule's **root-mean-square** or **rms** speed (which, you've got to admit, is easier to say than "the square root of the average of the squared speed" of the molecule). The rms speed for helium atoms at room temperature (295 K) turns out to be about 1400 m/s, whereas the rms speed for nitrogen molecules is about 520 m/s. Note that the speed of sound (at 340 m/s) is about the same order of magnitude. This makes sense: since air molecules are not physically in contact with each other most of the time and do not influence each other at a distance, the only way that information about a disturbance in the air can move from one point to another nearby is if molecules physically travel the distance between those points.

The equipartition theorem applies even in the case where the ratio m_2/m_1 between molecules in the mixture is extremely large. In fact, it is true even in the case where the "molecules" are actually tiny but visible particles of matter. In 1827, the English botanist Robert Brown discovered that grains of pollen suspended in water seemed to move continuously and randomly. At first, this motion (which came to be called **Brownian motion**) were taken as a sign of life, but careful experiments showed that small inorganic particles behaved similarly.

This behavior was unexplained until the equipartition theorem was developed. In 1905, Albert Einstein (in the same issue of *Annalen der Physik* that contained his first paper on special relativity) published a detailed quantitative description of Brownian motion based on the equipartition principle. Einstein argued that the motion that Brown had observed was due to collisions between water molecules and the pollen grains, and that the pollen grains were exhibiting the random motion to be expected from the equipartition theorem.

Moreover, Einstein noted, it should be possible to determine Avogadro's number by studying Brownian motion. If molecules are extremely small (which is what happens if Avogadro's number is large), then many molecules will strike the grain from many directions at once during a given time interval, and the forces that they exert will mostly cancel out. On the other hand, if molecules are few in number and large in mass, the particles striking the grain at a given time will more likely exert unbalanced forces, leading to greater jostling of the grain. Einstein derived a quantitative relation between a grain's mass, its average displacement during jostling, and Avogadro's number, making it possible to deduce Avogadro's number from measurements of Brownian motion.

In 1908, Jean Perrin published results from a brilliant series of experiments using this and other aspects of Brownian motion to compute Avogadro's number, arriving at a figure of 6×10^{23} molecules per mole. His work, which earned him the 1926 Nobel Prize in physics, convinced the skeptics in the physics community that atoms were real objects, not simply mathematical conveniences. These results also proved wrong the critics of Boltzmann's work in statistical mechanics (sadly, they were published two years after Boltzmann's death).

Einstein's method for finding Avogadro's number

Exercise T2X.7: Show that the rms speed of helium at 295 K is indeed about 1400 m/s and the rms speed of nitrogen is about 520 m/s. (Remember that helium has a mass of about 4 g per mole and nitrogen has a mass of about 28 g per mole, and 1 mole = 6.02×10^{23} molecules.)

T2.6 SOLIDS AND LIQUIDS

The same general principles apply to liquids and solids as well. The main difference between liquids and gases is that the molecules in a liquid, while somewhat free to move, are close enough to touch, making the liquid comparatively incompressible. The fact that the molecules in a liquid are almost always in contact means several of the assumptions we used in "deriving" the ideal gas law (specifically, the assumptions that each molecule can be treated in isolation and do not significantly interact) do not apply. The significant but complex interactions between molecules makes it hard to construct a microscopic model of liquids having the quantitative accuracy of the ideal gas model. (The equipartition theorem also breaks down for liquids because we cannot write an independent energy equation for each molecule). Still, the most general principles that apply to gases apply to liquids as well: liquids do store thermal energy (on the order of a few times $k_B T$ per molecule) in the form of kinetic energy of molecular motion and the potential energy of molecular interactions.

In the case of solids, these intermolecular interactions become so strong that the molecules are essentially locked into fixed locations in a lattice. A reasonably successful microscopic model for a solid is to imagine that each molecule is independent and held in its lattice position as if by springs as shown in Figure T2.5: the springs qualitatively represent the forces exerted on the molecule by its neighbors in the lattice. This **bedspring model** is an idealization of a solid in much the same way that the ideal gas model is an idealization of a gas.

To the extent this model is true, we can (as in the case of a gas) write an expression for the total energy contained by each individual molecule so that we can apply the equipartition theorem. The expression in this case is

$$E_{\text{tot}} = \tfrac{1}{2} m v_x^2 + \tfrac{1}{2} m v_y^2 + \tfrac{1}{2} m v_z^2 + \tfrac{1}{2} k_s x^2 + \tfrac{1}{2} k_s y^2 + \tfrac{1}{2} k_s z^2 \qquad \text{(T2.21)}$$

where m is the molecule's mass, k_s is the effective spring constant of the interactions holding the molecule in its lattice position, and the velocity and displacement components are measured relative to its equilibrium position. Since there

No easy models for liquids

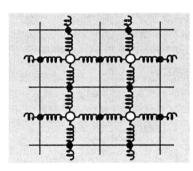

Figure T2.5: Bedspring model for a solid. In this model, each molecule (white ball) is imagined to be held in its lattice position by springs. The diagram shows a single two-dimensional layer of such a solid: you should imagine many such layers stacked above and below the plane of the drawing, with springs running perpendicular to the drawing as well.

Predicted relationship between *U* and *T* for a solid

are six terms here, the equipartition theorem predicts that each molecule has an average energy of $\frac{6}{2}k_BT$ and that an *N*-atom solid has a total internal energy of

$$U = \tfrac{6}{2}Nk_BT = 3Nk_BT \qquad\qquad (T2.22)$$

Measured values of *dU/dT* for many solids at normal temperatures are fairly consistent with this prediction.

SUMMARY

I. THE IDEAL GAS LAW
   A. Empirical observations of gases led to the *ideal gas law*: $PV = Nk_BT$
      1. where *P, V, T* are pressure, volume, and temperature respectively
      2. *N* is the number of molecules in the sample of gas
      3. k_B is Boltzmann's constant = 1.38×10^{-23} J/K for all gases
   B. The chemist's version of the law is $PV = nRT$
      1. n = number of moles = $N/N_A = M/M_A$
      2. N_A = Avogadro's number, M_A = mass in grams per mole

II. A MICROSCOPIC MODEL OF AN IDEAL GAS
   A. Assumptions of the model:
      1. *N* is huge
      2. The gas molecules are tiny compared to space they occupy
      3. Molecules behave like newtonian particles
      4. There are no interactions between gas molecules
      5. All the molecules in the gas are identical
      6. All collisions with the walls are elastic
      7. The motion of molecules in the gas is entirely random
   B. The ideal gas law emerges from this model
      1. Thinking about momentum transfer to walls $\Rightarrow PV = Nm[v_x^2]_{\text{avg}}$
      2. This becomes the ideal gas law if we identify $\frac{1}{2}m[v_x^2]_{\text{avg}} = \frac{1}{2}k_BT$
      3. Since $[v_x^2]_{\text{avg}} = [v_y^2]_{\text{avg}} = [v_z^2]_{\text{avg}}$, this means that $K_{\text{avg}} = \frac{3}{2}k_BT$

III. THE EQUIPARTITION THEOREM
   A. Background for the theorem
      1. The expression for the energy of an isolated molecule involves a sum of terms quadratic in a velocity or displacement component
      2. Each such term describes a molecular *degree of freedom*
      3. There are typically three such terms associated with CM motion, three associated with rotation, two for each mode of vibration, etc.
   B. Statement of the theorem: The molecular energy storage mode associated with each degree of freedom gets an average of $\frac{1}{2}k_BT$ of energy
   C. Limitations imposed by quantum mechanics:
      1. degrees of freedom corresponding to rotations that leave the molecule completely unchanged do not count
      2. If differences between a storage mode's energy levels are $>> k_BT$, then a typical collision cannot excite the mode out of its ground state and thus cannot transfer energy to it: the mode is *frozen out*
      3. A rule of thumb: mode is fully unfrozen when $k_BT > 2\Delta E$
   D. Application of these ideas to gases
      1. Monatomic gases *only* have translational modes, so $U = \frac{3}{2}Nk_BT$
      2. Diatomic gases have 3 translational and 2 rotational modes, but vibration is frozen out at normal temperatures, so $U = \frac{5}{2}Nk_BT$

IV. MIXTURES, LIQUIDS, AND SOLIDS
   A. Mixed gases
      1. Equipartition applies to each gas separately so $v_{\text{rms}} = [3k_BT/m]^{1/2}$
      2. It even applies to macroscopic particles $\Rightarrow$ Brownian motion
   B. Liquids: no simple model exists, but energy / molecule is still $\approx k_BT$
   C. Bedspring model for solids $\Rightarrow$ 6 degrees of freedom $\Rightarrow U = 3Nk_BT$

GLOSSARY

Boyle's law: the empirical statement that if the temperature and the number of molecules in a gas are fixed, the pressure of a gas is inversely proportional to its volume: PV = constant.

ideal gas law: an empirical law $PV = Nk_BT$ that links the pressure P, volume V, number of molecules N, and temperature T of a gas. This law is an idealization that all real gases approach in the limit of zero density.

Boltzmann's constant k_B: the physical constant $k_B = 1.38 \times 10^{-23}$ J/K $= 8.63 \times 10^{-5}$ eV/K appearing in the ideal gas law, but which more fundamentally links a molecule's average KE to the ambient temperature: $K_{avg} = \frac{3}{2}k_BT$.

gas constant R: the product of Boltzmann's constant and Avogadro's number: $R = k_BN_A = 8.31$ J·K^{-1}·mole^{-1}.

molecular mass M_A: the mass (in grams) of Avogadro's number (one mole) of molecules of a given substance. This mass is *roughly* equal to the number of protons and neutrons in the molecule's atomic nuclei.

diatomic (molecule): a molecule containing two atoms.

polyatomic (molecule): a molecule constructed of more than two atoms.

equipartition theorem: asserts that every molecule has an integer number of *degrees of freedom*, and that on the average, each of a molecule's degrees of freedom gets $\frac{1}{2}k_BT$ of energy (unless quantum limits squash it).

degree of freedom: We can usually write the expression for the total energy of a molecule as a sum of terms that have the form of a constant times the square of a velocity or displacement component. Each such term corresponds to a *degree of freedom* for that molecule.

quantum limits model puts two limits on the free application of the equipartition theorem. (1) The first limit asserts that a degree of freedom corresponding to a rotation that doesn't change the molecule (independent of the angle of rotation) doesn't count as a degree of freedom in the equipartition theorem. (2) The second limit asserts that a degree of freedom might be "frozen out" of the energy exchange assumed in the equipartition theorem if k_BT is much smaller than the difference ΔE between energy levels for the energy storage mode associated with that degree of freedom. When k_BT is larger than about $2\Delta E$, the mode participates almost fully, but if k_BT is less than $\frac{1}{2}\Delta E$, it is completely frozen out, as if the degree of freedom didn't exist.

root-mean-square (rms) speed v_{rms}: the square root of the average squared speed: $v_{rms} \equiv ([v^2]_{avg})^{1/2}$.

Brownian motion: the jiggling of small particles suspended in a solution due to impact of water molecules.

bedspring model: a simple model for a solid that imagines that each molecule is held in its lattice location by springs and vibrates independently from the other molecules. (Albert Einstein was one of the first to describe the thermal consequences of this model.)

TWO-MINUTE PROBLEMS

T2T.1 3×10^{23} molecules of H_2 gas at a temperature of 100 K are placed in a 1 m^3 container. What is the pressure?
A. 4.1 atm D. 300 Pa
B. 410 atm E. 3×10^{25} Pa
C. 4100 Pa F. other (specify)

T2T.2 The ideal gas model assumes that molecules have infinitesimal size. Consider one molecule bouncing around in the container. As the molecule's size increases relative to the size of the container, how will the average pressure P that it exerts change, other things being held constant?
A. P increases C. P decreases
B. P does not change D. more information needed

T2T.3 If the speed of a molecule doubles, the average pressure that it exerts on the wall of its container (other things remaining the same)
A. remains the same C. quadruples
B. doubles D. more information needed

T2T.4 The average x-velocity of molecules in a container of gas at rest is zero (T or F).

T2T.5 $[v_x]_{avg} = \sqrt{[v_x^2]_{avg}}$ (T or F).

T2T.6 Two identical rooms, A and B, are sealed except for an open doorway between them. Room A is warmer than B. Which room has the greater number of molecules?
A. Room A B. Room B
C. Both have the same number of molecules.

T2T.7 One mole of hydrogen gas and half a mole of helium gas are mixed together in a container and maintained at a fixed temperature T. How does the total pressure P_H exerted by the hydrogen gas compare to the total pressure P_{He} exerted by the helium molecules? Note that a molecule (atom) of helium has twice the mass of an H_2 molecule.
A. $P_H = 2P_{He}$ D. $P_H = \sqrt{\frac{1}{2}}P_{He}$
B. $P_H = \sqrt{2}P_{He}$ E. $P_H = \frac{1}{2}P_{He}$
C. $P_H = P_{He}$ F. other (specify)

T2T.8 Equal numbers of molecules of two diatomic gases with molecular masses m and M (with $M > m$) respectively are mixed in a large container and maintained at a fixed temperature. Do the molecules with the larger mass
a. have a greater (A), smaller (B) or the same (C) v_{rms}
b. have a greater (A), smaller (B) or the same (C) K_{avg}
c. store more (A), less (B), or the same (C) total energy
d. exert more (A), less (B), or the same (C) pressure
than the molecules with the smaller mass?

T2T.9 The difference between vibrational energy levels in a certain molecule is about 0.008 eV. At room temperature, this energy storage mode is A. completely frozen out
B. partially frozen out C. fully engaged

T2T.10 The difference between vibrational energy levels in a certain molecule is about 0.013 eV. This mode will be completely frozen out at temperatures below 150 K (T or F).

HOMEWORK PROBLEMS

BASIC SKILLS

T2B.1 In interstellar space, there is about one H_2 molecule per cubic centimeter. The temperature of deep space is about 3 K. What is the pressure of this interstellar gas? How does this compare to pressure of the best vacuum that can presently be attained in the laboratory ($\approx 10^{-13}$ Pa)?

T2B.2 A bottle of oxygen at room temperature (295 K) has a pressure of 2.0×10^5 Pa (about 2 atm) and occupies a volume of 0.01 m^3. How many molecules does it contain?

T2B.3 Consider a gas whose molecules have an average kinetic energy of 1 eV. What is the temperature of the gas?

T2B.4 Find the rms speed of CO_2 molecules in a room at room temperature (295 K). (Carbon atoms have 12 and oxygen atoms have 16 nucleons in their nuclei.)

T2B.5 The rms speed of Cl_2 molecules in a sample of gas is about 180 m/s. What is the temperature of the gas? (Chlorine atoms have 35 nucleons in their nuclei.)

T2B.6 At a temperature of 210 K, the vibrational modes of an ammonia molecule (NH_3) are frozen out, but its rotational modes are unfrozen. What would you estimate the value of dU/dT for ammonia at this temperature to be?

T2B.7 Imagine that the difference between vibrational energy levels in a certain molecule is 0.035 eV. Above roughly what temperature will this mode become unfrozen?

SYNTHETIC

T2S.1 In the early 1800's John Dalton first expressed the law of partial pressures: "the total pressure of a mixture of gases is equal to the sum of the partial pressures of the gases making up the mixture". Carefully explain how this law follows from the molecular model of gases.

T2S.2 What is the total thermal energy of 0.40 g of helium gas inside a balloon at room temperature? How fast would you have to run to have that much kinetic energy?

T2S.3 The average velocity of a molecule in a gas at rest is zero. How do we know this? The rms speed of a molecule in a gas at rest is not zero. How is this possible?

T2S.4 Explain why in the case of a diatomic or polyatomic gas, the quantity K_{avg} in equation T2.15 must refer to the average kinetic energy of the molecule's center of mass, *not* including its rotational and/or vibrational energy.

T2S.5 A mixture of an equal number of hydrogen and oxygen molecules are placed in a container and held at a fixed temperature. Which type of molecule has the greater rms speed and roughly how many times greater is that speed than the rms speed of the other type? Which type of gas

stores more energy as a whole? (H_2 has a molecular mass of 2 g/mole, while O_2 has a molecular mass of 32 g/mole.)

T2S.6 Find the rms speed of H_2 molecules in the upper atmosphere, where the temperature is nearly 1000 K. This is just an *average* speed: the speeds of individual molecules will be randomly spread about this value. The earth's escape velocity is about 11.2 km/s. Speculate about why the earth's atmosphere contains little hydrogen.

T2S.7 Imagine that we add 25 J of thermal energy to a liter (1000 cm^3) of air at room temperature and normal pressure. By about how much does its temperature increase? *Hint:* The relative amount of O_2 and N_2 in air is irrelevant. Why?

RICH-CONTEXT

T2R.1 Edward Purcell, a physicist at MIT, proposed the following problem as a way to learn about the magnitude of Avogadro's Number. A ___ of water contains about as many molecules as there are ___s of water in all of the oceans of the earth. What single word best fits the two blank spaces: drop, teaspoon, tablespoon, cup, quart, barrel or ton? Oceans cover about 70% of the earth's surface area and have an average depth of about 5 km.

ADVANCED

T2A.1 We can crudely estimate the rotational energy levels of a hydrogen molecule this way. Imagine observing the molecule from the frame where its center of mass is at rest, and let the momentum of each atom as it rotates around the center of mass be p. An atom with momentum p will have a wavelength $\lambda = h/p$ by the de Broglie relation. But the quantum wavefunction associated with the atom will have to fit an integer number of wavelengths along the atom's circular path, or the wave won't match itself after going around the circle once. If the atom is a distance r from the center of mass, the length of its circular path is $2\pi r$. The condition that the wave must match itself after going around the circle once means that $n\lambda = 2\pi r$, where n is some integer. Use this expression, the de Broglie relation, and $K = p^2/2m$ to derive an expression for the possible rotational kinetic energies of the molecule (don't forget that there are two atoms involved!). Estimate the energy difference between the ground state ($n = 0$) and the next higher rotational state ($n = 1$) if the distance between the molecules is equal to the diameter of the hydrogen atom (0.11 nm). You should get roughly 0.015 eV.

[*Note*: this "derivation" involves a patchwork of quantum and classical ideas, and is not at all rigorous. Still, it happens to yield a reasonable estimate of the desired result, as a more careful derivation verifies.]

ANSWERS TO EXERCISES

T2X.1 3.47×10^{22} molecules, 1.62 g.

T2X.2 Let dp be the momentum transferred per collision, and let dt be the time interval between collisions. Then we have $F_x \equiv dp/dt = +2mv_x/(2L/v_x) = mv_x^2/L$

T2X.3 Note that $\frac{1}{2}[v_x^2 + v_y^2 + v_z^2]_{avg} = \frac{1}{2}m[v_x^2]_{avg} + \frac{1}{2}m[v_y^2]_{avg} + \frac{1}{2}m[v_z^2]_{avg}$, since the average of a sum is the sum of averages. Now T2.14 and T2.13 imply T2.15.

T2X.4 $\frac{5}{2}N_Ak_B = 20.8$ J/K, $\frac{6}{2}N_Ak_B = 24.9$ J/K.

T2X.5 The answer is given (this is a simple calculation).

T2X.6 Since the lowest rotational energy levels for H_2 are about 0.015 eV apart, the rotational modes will be frozen out when k_BT is below about 0.0075 eV $\Rightarrow T \approx 90$ K.

T2X.7 The answer is given. (*Hint*: Values of m for He and N_2 are 6.64×10^{-27} kg and 4.65×10^{-26} kg respectively.)

T3

GAS PROCESSES

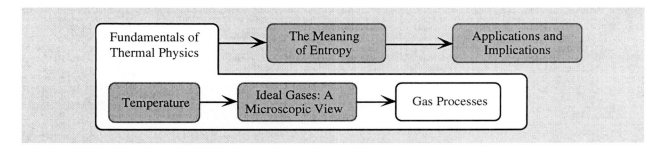

T3.1 OVERVIEW

In the last chapter, we examined the question of how thermal energy was related to temperature in an ideal gas. We eventually came to see that the thermal energy in any substance is the sum of the kinetic, rotational, vibrational energies of its molecules (and the potential energies of their interactions), and that these energies are linked to the temperature.

In this chapter, we will look at quantitative applications of the ideal gas model, exploring what happens to the thermal energy of a gas when we compress it or allow it to expand. Understanding such processes will be essential preparation for our examination of heat engines at the end of the unit. Here is overview of the sections of this chapter.

T3.2 *REVIEW OF HEAT AND WORK* reviews the distinction between heat and work presented way back in Chapter C9 of unit *C*.

T3.3 *WORK DURING EXPANSION OR COMPRESSION* explains why expansions and compressions change the thermal energy in a gas.

T3.4 *PV DIAGRAMS AND CONSTRAINED PROCESSES* shows how we can use *PV* diagrams to illustrate compression and expansion processes and defines some terms useful for describing such processes.

T3.5 *COMPUTING THE WORK* discusses how we can actually compute the work energy flow during an expansion or compression.

T3.6 *ADIABATIC PROCESSES* examines in depth the important special case of processes where a gas changes volume but cannot exchange heat energy with its surroundings.

T3.2 REVIEW OF HEAT AND WORK

We introduced the concept of heat and work in Chapter C9 of Unit *C*. The main point in that chapter was to introduce and emphasize the distinctions between *thermal energy, heat,* and *temperature.* In this chapter, we will focus more on the distinction between *heat* and *work.* A brief review of these important technical terms is thus in order.

In thermal physics we are often interested less in an object's total thermal energy than in how that energy changes under certain conditions. Knowing how that energy changes involves watching the amount of energy that crosses the object's boundaries. The crucial thermodynamic concepts *heat* and *work* therefore are both defined to describe *energy transfer across a system boundary.*

Definition of *heat*

When a hot object is placed in contact with a cold object, energy spontaneously flows across the boundary between them until both objects come to have the same temperature. As a result, the thermal energy of the hot object decreases and the thermal energy of the cold object increases. In physics, **heat** is any energy that crosses the boundary between the two objects *because* of a temperature difference across the boundary. Let me emphasize that to be *heat,* the energy in question *must*

1. be flowing across some kind of boundary between systems,
2. as a direct result of a temperature difference across that boundary.

Definition of *work*

We define **work** in thermal physics to be any *other* kind of energy flowing across the boundary of a system. For example, if I stir a cup of water vigorously and it gets warm as a result, I have not "heated" the water, I have done *work* on it: the mechanical energy flows across the boundary of the water not because of a temperature difference but because of my stirring effort.

Note that *heat* and *work* both refer to energy in transit across a boundary. This sharply distinguishes both from *thermal energy,* which refers to energy *inside* the system boundary. Both heat and work flows can contribute to changes in the thermal energy. In fact, conservation of energy implies that:

The first law of thermodynamics

$$\Delta U = Q + W \qquad\qquad (T3.1)$$

where ΔU is the change in a system's thermal energy in a given process, Q is the heat energy added in the process, and W is the work energy that has flowed into the system in the process. We call this crucial equation the **first law of thermodynamics.** The definitions of heat and work are illustrated in Figure T3.1a. (Note that in this text, we will consider Q and W to be negative if energy flows *out* of the system: some other books follow a convention where the sign of W is defined the other way.)

Distinction between heat and work can depend on choice of boundary

The distinction between heat and work is more a matter of human definition and convenience than of deep physics: at the microscopic level, an energy flow is simply an energy flow. Moreover, whether we call a flow of energy in a given process "heat" or "work" can depend on one's choice of system boundary. For example, consider the situation shown in Figure T3.1b. Electricity flows into a piece of nichrome wire that is immersed in a cup of water. The electricity causes the temperature of the wire to increase, which in turn causes the temperature of the water to increase. If we draw the boundary labeled *a* (making "the system" the water and the cup and the resistor), then the energy crossing the boundary is electrical energy and thus *work.* If one draws the boundary labeled *b* (making "the system" the water alone), then the energy crossing the boundary is *heat,* driven by the temperature difference between the wire and the water.

You will become confused by much of what follows if you don't understand the technical distinctions between *heat, work,* and *thermal energy.* It is particularly easy to misuse the word heat (even some physics texts do!). The following exercise should help develop your ability to distinguish *heat* and *work.*

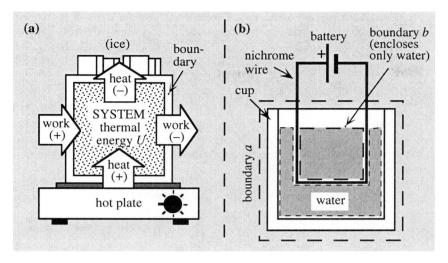

Figure T3.1: (a) Illustration of the definitions of heat and work. **(b)** Whether an energy flow in a given situation is *heat* or *work* can depend on how we define the system boundary. If we define the system to be enclosed by boundary *a*, the energy flowing into the system is *work*. If we define the system to be enclosed by boundary *b*, the energy flowing into the system is *heat*.

Exercise T3X.1: In each process described below, energy flows from object *A* to object *B*. Is the energy flow involved *heat* or *work*?   heat or work?

a. The soup (*B*) in a pan sitting on an electric stove (*A*) gets hot.   ____

b. Light from an incandescent bulb (*A*) flows to its surroundings (*B*)   ____

c. You (*A*) compress air (*B*) in a bike pump, making it warm.   ____

d. Your hands (*A*) are warmed when they face a fire (*B*).   ____

e. The atmosphere (*A*) warms a re-entering spacecraft (*A*).   ____

f. A hot pie (*A*) becomes cooler sitting in the kitchen (*B*).   ____

g. Your chair (*B*) becomes warmer after you (*A*) sit in it a while.   ____

h. A drill bit (*B*) becomes hot after being spun by the drill (*A*).   ____

T3.3 WORK DURING EXPANSION OR COMPRESSION

Work can be done on or by a system in a variety of ways, but in much of this unit we will be primarily concerned with work done as a result of the compression or expansion of a gas. Understanding how work is involved in expansion and compression will be especially important to us at the end of the unit when we discuss how heat can be converted to mechanical energy using a **heat engine** (such as a gasoline or steam engine). Most heat engines use expansion and compression of some gas to perform this conversion.

Imagine a gas confined to a cylinder by a piston, and imagine that an external interaction exerts a force $\vec{F}_{ext}$ on the piston that pushes it *slowly* inward an infinitesimal distance dx (see Figure T3.2). As it pushes the piston inward, the amount of energy that the interaction transfers to the piston is

$$dK = F_{ext,x}dx = F_{ext}|dx| \qquad (T3.2)$$

since $\vec{F}_{ext}$ acts entirely in the $-x$ direction here, and dx is negative in this compression process. Now, the gas in the cylinder exerts an opposing force $\vec{F}_{gas}$ against the piston. If we push the piston *slowly* so that its speed remains roughly constant (implying that $F_{gas} \approx F_{ext}$), then the piston's interaction with the gas *extracts* energy at the same rate that the external interaction supplies it. Where does this energy go? It must go to thermal energy in the gas. Since this energy flow is not due to a temperature difference, it is *work*. Therefore, the infinitesimal work done during such an infinitesimal slow compression is

$$dW = F_{ext}|dx| = F_{gas}|dx| \qquad (T3.3)$$

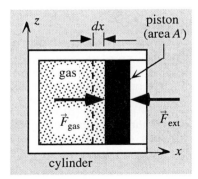

Figure T3.2: Slowly compressing gas in a cylinder.

Now, the magnitude of the force $\vec{F}_{gas}$ exerted by the gas on the piston in this process is simply the pressure exerted by the gas times the area of the piston:

$$F_{gas} = PA \qquad\qquad (T3.4)$$

Combining equations T3.4 and T3.3, we see that the infinitesimal amount of work done on the gas by this compression process is given by:

$$dW = PA|dx| \qquad \text{(during slow compression)} \qquad (T3.5)$$

We can simplify this expression by noting that the infinitesimal change in the volume of the gas during this process is:

$$dV \equiv V_{final} - V_{initial} = -A|dx| \qquad \text{(for compression)} \qquad (T3.6)$$

The negative sign is necessary because during a compression, $V_{final} < V_{initial}$, so dV is negative. Plugging this into equation T3.4, we get:

Expression for infinitesimal work involved in a slow compression or expansion

$$dW = -PdV \qquad \text{(for any slow volume change)} \qquad (T3.7)$$

In this equation, the negative sign ensures that dW is positive (saying that energy is flowing into the gas) for a compression even though dV is negative.

This formula is correct for an infinitesimal slow *expansion* as well! In the case of expansion, $dV = +A|dx|$ (positive) but $dW = -PA|dx|$ (negative) since the external force *extracts* kinetic energy from the piston as the piston moves outward against the force, so the gas must *supply* kinetic energy (at the expense of thermal energy). Combining these equations yields equation T3.7 again (in this case the negative sign ensures that dW is negative when dV is positive).

Exercise T3X.2: Equation T3.7 strictly applies only in the case where the compression is infinitesimal. Why? What generally happens to the pressure of a gas when it is compressed?

A microscopic model for how energy is transformed

How exactly is energy being transferred to the gas in the case of a compression, or from the gas in the case of an expansion? We can understand the microscopic processes involved in such an expansion using a simple analogy. Imagine bouncing tennis balls off the back end of a truck (see Figure T3.3). If the truck is stationary and the balls are perfectly elastic, they will bounce back with about the same kinetic energy as they had before striking the truck. But if the truck is backing up toward you, the balls will bounce back from the truck with more energy relative to you than you gave them: some of the energy of the truck's motion is being converted to kinetic energy in the balls. Similarly, if the truck is going away from you, the balls will bounce back with less energy than they had to begin with: energy is being transferred from the balls to the truck.

Figure T3.3: A ball thrown against the back of a truck will bounce back with **(a)** about the same energy as it had originally if the truck is at rest, **(b)** more energy if the truck is backing up, **(c)** less energy if the truck is moving forward.

In a similar way, gas molecules that bounce back elastically from a piston at rest will bounce back from an advancing piston (that is, during compression) with more energy than they had originally, implying that the internal energy of the gas increases at the expense of the piston's energy. When the piston retreats (during expansion), the molecules bounce back with less energy than they had originally, transferring the thermal energy of the gas to the motion of the piston.

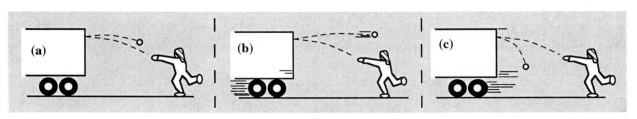

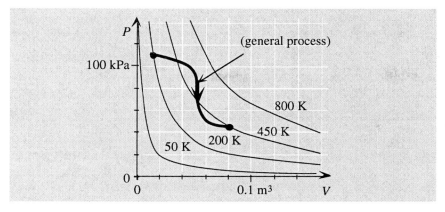

Figure T3.4: Every point on this *PV* diagram represents a possible macroscopic state for a fixed amount of ideal gas. The thin curves connect gas states having the same temperature (the curves are labeled assuming that $N = 5.8 \times 10^{23}$). Any quasistatic process is an ordered succession of states, and can be represented by a succession of points (that is, a curve) on this diagram.

T3.4 *PV* DIAGRAMS AND CONSTRAINED PROCESSES

How we describe the macroscopic state of a gas

We can describe the **macroscopic state** of an ideal gas in equilibrium by specifying the type of gas and values for four macroscopically measurable quantities: the pressure P of the gas, its volume V, its temperature T, and the number of molecules N involved. All other macroscopic properties of the gas can be computed from these. For example, the *mass* of the gas can be found knowing N and the type of gas, an ideal gas' thermal energy is a function of T, N, and the type of gas ($U = \frac{3}{2} N k_B T$ for a monatomic gas), and so on.

The quantities P, V, T and N are not independent: they are linked by the ideal gas law $PV = N k_B T$. Thus we really only need to know *three* of these variables to specify the state of a given type of the gas completely. Moreover, the number of gas molecules in a container is usually fixed, so the state of a fixed amount of gas is completely determined by *two* independent variables.

Representing states and processes on a *PV* diagram

If we choose these variables to be P and V, then the state of a fixed amount of gas can be represented by a point on a ***PV* diagram**, which is a simple plot of P versus V, such as the one shown in Figure T3.4.

Since $PV = N k_B T$, all of the points on a *PV* diagram corresponding to a given temperature (assuming N is fixed) lie on a curve such that $PV = $ (const.) or $P \propto 1/V$. A set of such curves for various values of T are shown in Figure T3.4.

A **quasistatic gas process** is defined to be any process involving gas that is done slowly enough that the gas remains essentially in equilibrium at all times. This means that at all times the gas pressure P has the *same* well-defined value at all points in the gas. As the state of the gas changes during the process, the point representing its state on the *PV* diagram will move, marking out a *path* on the diagram. An example path is shown in Figure T3.4.

Important (and useful) gas processes

P and V are completely independent variables, and there is no intrinsic reason why a gas can't have any pressure at any given volume. A process might in principle (like the "general process" shown in the figure) involve any connected sequence of points on the diagram. In practice, though, there are four special gas processes that keep coming up as useful approximations in realistic situations:

1. **isochoric processes** where the gas is heated or cooled while its volume is constrained to be constant (for example, by keeping the gas in a rigid container). The root *iso-* means "same" and *choric* refers to volume.

2. **isobaric processes** where the gas is heated or cooled while its *pressure* is constrained to be constant (for example, by confining the gas with a piston whose other side is acted on by the constant pressure of the earth's atmosphere. The root *bar-* refers to pressure (as in "barometer").

3. **isothermal processes** where the gas is expanded or compressed while its temperature is constrained to be constant (for example, by placing it in good thermal contact with a "bath" having a certain fixed temperature).

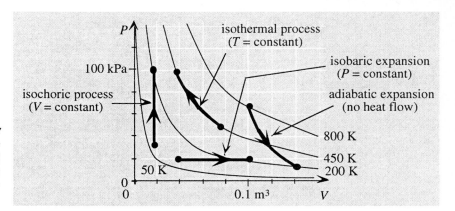

Figure T3.5: Curves on a *PV* diagram corresponding to the four most useful types of constrained processes. Again, the values on the curves of constant temperature assume that $N = 5.8 \times 10^{23}$.

4. **adiabatic processes** where the gas is expanded or compressed while heat flow to or from the gas is constrained to be zero (for example, by putting thermal insulation around the gas container). We'll discuss adiabatic processes more fully in section T3.6.

These four processes (see Figure T3.5) are called **constrained processes**, because the particular path on the *PV* diagram that each process follows is determined by a *constraint* placed on the gas during the process (for example, the gas in an isothermal process is constrained to have a constant temperature).

T3.5 COMPUTING THE WORK

How we find *W* when the pressure changes significantly during the process

These constrained processes are useful partly because we can actually compute the work done on or by the gas involved in such a process. As we noted before, equation T3.7 ($dW = -PdV$) applies only to infinitesimal compressions or expansions. This is because as a gas is compressed or expanded, its pressure will normally change, meaning that the value of *P* will be different for different parts of the process. In general, if we want to compute the total work *W* done during a significant compression or expansion process, we have to do an integral:

$$W = -\int PdV \qquad (T3.8)$$

In words, this equation tells us that to compute the total work, we should divide the process into infinitesimal steps, compute $dW = -PdV$ for each step, and sum the result over all the steps.

Implications for particular constrained processes

In order to actually evaluate this integral, we have to be able to express *P* during the process as a function of *V*. This is easy to do for each of the four special constrained processes listed on the previous page except for the adiabatic process, which we'll consider in the next section. The *isochoric* process (where *V* is constant) is trivial: since the volume does not change ($dV = 0$), no work is done as the result of compression or expansion:

$$W = 0 \qquad (T3.9a)$$

In an *isobaric* process, *P* is constant, so we can pull it out of the integral:

$$W = -\int PdV = -P\int_{V_i}^{V_f} dV = -P(V_f - V_i) = -P\Delta V \qquad (T3.9b)$$

where V_i is the gas' initial volume and V_f is its final volume.

In an isothermal process, *T* is constant. The ideal gas law then implies that:

$$PV = Nk_BT = \text{constant} = P_iV_i \quad \Rightarrow \quad P(V) = \frac{Nk_BT}{V} = \frac{P_iV_i}{V} \qquad (T3.10a)$$

where P_i and V_i are the gas' initial pressure and volume respectively, and $P(V)$ means "pressure as a function of volume". Plugging this into equation T3.8, using $\int x^{-1}dx = \ln x$ and $\ln x - \ln y = \ln(x/y)$, you should be able to show that

$$W = -Nk_BT\ln\left(\frac{V_f}{V_i}\right) = -P_iV_i\ln\left(\frac{V_f}{V_i}\right) \tag{T3.10b}$$

Exercise T3X.3: Verify this.

Exercise T3X.4: A gas initially at atmospheric pressure (100 kPa) in a box 10 cm on a side is isothermally compressed to half that volume. What is W?

One of the reasons that PV diagrams of gas processes are so useful is that *the work done in a given quasistatic expansion or compression process is equal in magnitude to the area under the curve representing that process on a PV diagram*. This is a direct consequence of equation T3.8: if we consider P to be a function of V, then the standard interpretation of the integral of $P(V)$ is that it corresponds to the area of the curve of P when it is plotted as a function of V, (see Figure T3.6.)

Because of this simple visual interpretation of equation T3.8, one can get a lot of qualitative information about a process from a PV diagram, as demonstrated in the following example.

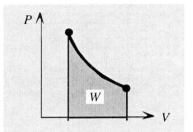

Figure T3.6: The *magnitude* of the work flowing into or out of a gas is equal to the area under the curve of P plotted as a function of V. The sign of the work is positive if the gas is being compressed, negative if it expands.

EXAMPLE T3.1

Problem: Imagine that an ideal monatomic gas is taken from its initial state A to state B by an *isothermal* process, from B to C by an *isobaric* process, and from C back to its initial state A by an *isochoric* process, as shown in Figure T3.7a. Fill in the signs of Q, W, and ΔU (or zero if appropriate) for each step (assuming that no work is done through expansion or compression).

Solution The finished chart is shown in Figure T3.7b. These signs are determined as follows.

Process $A \rightarrow B$ is an isothermal expansion, meaning that T is constant. But for an ideal gas, U is proportional to T ($U = \frac{3}{2}Nk_BT$ for a monatomic gas), so ΔU is *zero* during an isothermal process. During an expansion, work energy flows out of the gas (remember that for each infinitesimal step along this process, $dW = -PdV$, and $dV > 0$ in an expansion process), so W is *negative*. But if work energy flows out of the gas, and yet the gas' total thermal energy U remains the same, then (in the absence of other kinds of work), heat energy must flow into the gas, so Q must be *positive*.

Process $B \rightarrow C$ is an isobaric *compression*, so W is *positive* here. On the other hand, the ideal gas law says that $PV = Nk_BT$. Therefore, since V is decreasing in this process while P remains constant, T must be decreasing, (that is $dT < 0$ for each small step in the process) which in turn implies that the gas' thermal energy U must be decreasing: thus ΔU is *negative*. Since work energy is flowing *into* the gas, and yet its thermal energy is *decreasing*, heat energy must be flowing *out* of the gas: Q is negative.

Finally, process $C \rightarrow A$ is an isochoric process. Since there is no change in volume, no work is done on or by the gas: $W = 0$. Yet the temperature is increasing (since PV is increasing), so the gas' thermal energy is increasing, meaning that ΔU is *positive* in this process. This increase in thermal energy must be supplied by heat (since $W = 0$), so Q is positive.

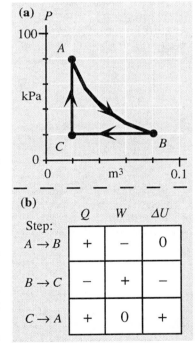

Figure T3.7: A sequence of gas processes.

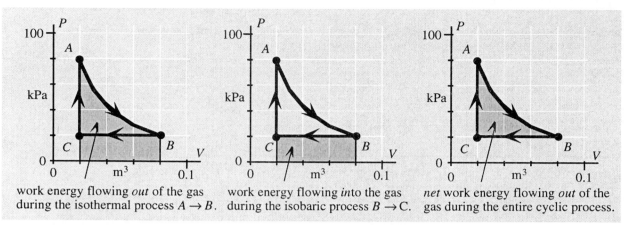

work energy flowing *out* of the gas during the isothermal process $A \rightarrow B$.   work energy flowing *into* the gas during the isobaric process $B \rightarrow C$.   *net* work energy flowing *out* of the gas during the entire cyclic process.

Figure T3.8: Net work energy flow during a cyclic process.

Work as the area under the curve on a *PV* diagram

We can also estimate directly from the diagram the work energy flowing into or out of the gas in the cyclic process we've been considering. Each grid square's worth of area on the *PV* diagram shown in Figure T3.8 represents:

$$(20 \times 10^3 \text{ Pa})(0.020 \text{ m}^3)\left(\frac{1 \text{ N/m}^2}{1 \text{ Pa}} \right)\left(\frac{1 \text{ J}}{1 \text{ N} \cdot \text{m}} \right) = 400 \text{ J} \qquad \text{(T3.11)}$$

There are a total of about 6 squares (four whole squares, two squares mostly complete, and two small parts of squares) of area under the curve for the isothermal expansion $A \rightarrow B$, so the gas loses $6 \times 400 = 2400$ J of work energy during that process. In the isobaric compression $B \rightarrow C$, the gas gains $3 \times 400 = 1200$ J of work energy. Since $W = 0$ for the isochoric process, the total energy lost by the gas during the entire cyclic process is about 1200 J. Note that *the net work done by the gas in a cyclic process is equal to the area enclosed by the process*.

Note that since the gas comes back to the same state *A* (and thus same temperature) at the end of the process that it had originally, its thermal energy must be the same at the end of the cycle as it was in the beginning (since U for an ideal gas only depends on *N* and *T*). Yet the gas loses about 1200 J of work energy in the cycle, as we've just seen. Where does this energy come from if not from the thermal energy of the gas? It must come from the *heat* energy that we put into the gas. This cycle is therefore an example of a process that converts heat energy into work energy. Many kinds of heat engines use such cyclic processes to produce useful work energy from heat energy. We'll talk more about heat engines in chapters T8 and T9.

Exercise T3X.5: The figure below shows another cyclic gas process. Fill out the chart as we did in example T3.1. Also estimate the net amount of work energy flowing out of the gas in this process. In which step of the process was energy supplied in the form of heat?

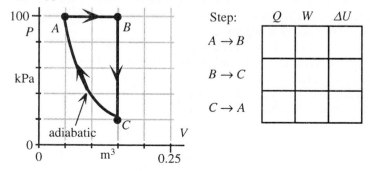

Step:	Q	W	ΔU
$A \rightarrow B$			
$B \rightarrow C$			
$C \rightarrow A$			

Exercise T3X.6: Check my estimate for the work energy lost during process $A \rightarrow B$ in Figure T3.8 using the exact formula T3.10b. Is our estimate within about 10% or so of the exact answer? (10% is about the best one can do by eye.)

T3.6 ADIABATIC PROCESSES

The last exercise included an *adiabatic* compression process as one of the steps in the cyclic process. In an adiabatic process, no heat is permitted to enter or leave the system in question (*adiabatic* is from Greek *adiabatos* "impassable", meaning the boundary is impassable to heat.)

The relationship between the pressure and volume of an ideal gas in an isothermal process (where $\Delta T = 0$) is given by the ideal gas law:

$$PV = Nk_BT = \text{constant} = P_iV_i \qquad \text{(isothermal process)} \qquad \text{(T3.12)}$$

where P_i and V_i are the gas' initial pressure and volume respectively (note that since PV is constant, this product will always have the same value that it had initially). In an adiabatic process, it turns out that the relationship between the pressure and volume of the gas is:

$$PV^\gamma = \text{constant} = P_iV_i^\gamma \qquad \text{(adiabatic process)} \qquad \text{(T3.13a)}$$

where: $\gamma = 1 + 2/n$, where n = degrees of freedom in gas molecule (T3.13b)

$\gamma = 1 + 2/3 = 1.67$ (monatomic gases) (T3.13c)
$\gamma = 1 + 2/5 = 1.40$ (diatomic gases at normal temperatures) (T3.13d)
$\gamma \approx 1 + 1/3 = 1.33$ (polyatomic gases at normal temperatures) (T3.13e)

How *P* and *V* are related in an adiabatic process

Equation T3.13a is a consequence of the ideal gas law, the fact that changes in the thermal energy of an ideal gas is given by $dU = \frac{n}{2}Nk_BdT$, and the fact that any work energy that comes out of the gas in an expansion comes at the expense of the gas' thermal energy: we'll work this out in a bit.

But before we dig into the derivation, let's see why the formula at least *qualitatively* makes sense. Note that $\gamma > 1$ for all types of gases. This means that if V doubles (for example), then P will have to decrease by a factor of $2^\gamma > 2$ to keep PV^γ constant. This means that the pressure falls off *more* sharply with increasing volume in an adiabatic process than it would in an isothermal process. Similarly, during adiabatic compression, the pressure of a gas rises more rapidly than it would in the isothermal case.

How does this make sense? In an adiabatic expansion, the work flowing out of the gas to its surroundings comes at the expense of its thermal energy, which means that *the gas' temperature will decrease.* (By contrast, in an isothermal expansion, heat energy has to be added to the gas to replace the work done by the gas, so that the internal energy and thus the temperature remains fixed.) Because the temperature of a gas falls as it adiabatically expands, its pressure decreases *more* than inversely with the volume (see Exercise T3X.7), instead of simply inversely with the volume (as it would if the gas' temperature were held constant). Similarly, when a gas is adiabatically compressed, its temperature increases, and thus its pressure increases *more* than inversely with the volume.

How to understand this relationship between *P* and *V*

Exercise T3X.7: Use the ideal gas law to explain qualitatively why the gas' decrease in temperature during an adiabatic expansion implies that the gas' pressure will fall more rapidly than it would during an isothermal expansion.

Adiabatic expansion or compression is an idealization that often very closely approximates real gas expansion or compression processes, especially in cases where the compression or expansion takes place so rapidly that there is no *time* for a significant amount of heat to flow in or out of the gas. For example, when you use a pump to inflate a bicycle tire, the gas in the pump gets about as hot as it would in a true adiabatic compression. It is not that heat *cannot* flow out of the gas (in fact, some does, making the pump barrel hot to the touch), it is that as you pump, there is insufficient time for a *significant* amount of heat to flow out of the gas as it is compressed.

Derivation of the adiabatic formula linking *P* and *V*

Now that we see how the adiabatic expansion formula works, let's see if we can derive it. We start by taking the derivative with respect to temperature *T* of both sides of the ideal gas law. Using the product rule, we get

$$P\frac{dV}{dT} + V\frac{dP}{dT} = Nk_B \quad \Rightarrow \quad PdV + VdP = Nk_B dT \qquad \text{(T3.14)}$$

This statement actually applies to any ideal gas process. What we'd like to do is to use something particular about the adiabatic process to eliminate all references to *T*, since our final formula does not involve *T*. For an infinitesimal adiabatic expansion, conservation of energy implies:

$$dU = dQ + dW = 0 + W = -PdV \qquad \text{(T3.15)}$$

since $Q = 0$ for an adiabatic process by definition. If we assume that in the temperature range in question, each gas molecule has *n* unfrozen degrees of freedom, then the equipartition theorem tells us that $dU = \frac{n}{2}Nk_B dT$. Plugging this into equation T3.15, and manipulating things a bit, we get:

$$Nk_B dT = -\frac{2}{n}PdV \qquad \text{(T3.16)}$$

Exercise T3X.8: Verify this.

If we plug this into equation T3.14, we can eliminate the $Nk_B dT$ term, and after a certain bit of algebraic manipulation, we get

$$\frac{dP}{P} = -\gamma\frac{dV}{V} \quad \text{where} \quad \gamma \equiv \left(1 + \frac{2}{n}\right) \qquad \text{(T3.17)}$$

Exercise T3X.9: Verify this.

Integrating both sides of equation T3.17 from its initial state (where it has pressure P_i and volume V_i) to whatever its current state is (where it has pressure *P* and volume *V*), we get:

$$\int_{P_i}^{P}\frac{dP}{P} = -\gamma\int_{V_i}^{V}\frac{dV}{V} \quad \Rightarrow \quad \ln P\big|_{P_i}^{P} = -\gamma\ln V\big|_{V_i}^{V}$$

$$\Rightarrow \quad \ln P - \ln P_i = -\gamma(\ln V - \ln V_i) \quad \Rightarrow \quad \ln\left(\frac{P}{P_i}\right) = -\gamma\ln\left(\frac{V}{V_i}\right) \quad \text{(T3.18a)}$$

Taking the exponential of both sides, we get

$$\frac{P}{P_i} = \left(\frac{V}{V_i}\right)^{-\gamma} = \left(\frac{V_i}{V}\right)^{\gamma} \quad \Rightarrow \quad PV^{\gamma} = P_i V_i^{\gamma} = \text{constant.} \qquad \text{(T3.18b)}$$

as previously claimed.

Problem: Imagine that we rapidly compress a sample of air whose initial pressure is 100 kPa and temperature is 22°C (= 295 K) to a volume that is a quarter of its original volume. What is its final temperature?

Solution This sudden compression is likely to be at least approximately adiabatic, so $P_f V_f^{\gamma} = P_i V_i^{\gamma}$. The ideal gas law says that $P = Nk_BT/V$. If we plug this into the first equation on both sides to eliminate P_f and P_i, divide by Nk_B and use $\gamma = 1.4$ for air (since almost all air molecules are diatomic), we get

$$T_f \frac{V_f^{\gamma}}{V_f} = T_i \frac{V_i^{\gamma}}{V_i} \quad \Rightarrow \quad T_f = T_i\left(\frac{V_i}{V_f}\right)^{\gamma-1} = (295 \text{ K})(4)^{0.4} = 514 \text{ K} \qquad \text{(T3.19)}$$

I. HEAT AND WORK
    A. We can keep track of a system's U by watching its boundaries
    B. Definition of *heat Q:* energy flowing across a system boundary that is driven by a temperature difference *alone.*
    C. Definition of *work W:* any *other* kind of energy flow across boundary
    D. The first law of thermodynamics: $\Delta U = Q + W$ (T3.1)
        1. This expresses conservation of energy in thermal context
        2. Q and W are defined to be positive if energy flows *into* the system
    E. The distinction between Q and W typically depends on exactly how we define our system's boundary

II. GAS PROCESSES IN GENERAL
    A. A *PV* diagram (a simple plot of P vs. V for a gas)
        1. The macroscopic *state* of a given gas is specified by $P, V, N,$ and T
        2. The ideal gas law actually links T to $P, V, N,$ and $U = (\text{const})Nk_BT$
        3. If N is fixed, then the state is specified by P and V alone
        4. The gas' state is thus represented by a *point* on a *PV* diagram
    B. *Quasistatic* processes
        1. If a process is quasistatic, P is always well-defined as V changes
        2. Therefore, the gas' state traces a well-defined curve on *PV* diagram
    C. *Constrained* processes follow specific kinds of curves on the diagram
        1. *isochoric:* $V = $ constant
        2. *isobaric:* $P = $ constant
        3. *isothermal:* $T = $ constant (so $P \propto 1/V$)
        4. *adiabatic:* $Q = 0$ (so $P \propto 1/V^{\gamma}$, as we'll see below)
    D. Computing the work during an expansion/compression process
        1. Basic principles of energy transfer imply that $dW = -PdV$ (T3.7) (Note that W is positive for compression, negative for expansion)
        2. When P changes during a process, we must integrate: $W = -\int PdV$ so magnitude of $W = $ area under the process' curve on *PV* diagram
        3. This means that the net work flowing out of a gas during closed cycle is equal to the area enclosed by the process on *PV* diagram
        4. Results of the integration for selected constrained processes
            a) isochoric: $W = 0$ (T3.9a)
            b) isobaric: $W = -P(V_f - V_i)$ (T3.9b)
            c) isothermal: $W = -Nk_BT\ln(V_f/V_i)$ (T3.10b)

III. ADIABATIC PROCESSES, defined to be processes where $Q = 0$
    A. This means that as a gas expands adiabatically and work flows out, U decreases, implying that T decreases. This means that the gas pressure P falls *faster* than $1/V$ as V increases (since $P = Nk_BT/V$).
    B. In fact, $PV^{\gamma} = $ constant $= P_i V_i^{\gamma}$, implying that $P \propto 1/V^{\gamma}$ (T3.13a)
        1. where $\gamma = 1 + 2/n$, $n = $ total number of degrees of freedom
        2. $\gamma = 1.67$ for monatomic, 1.4 for diatomic, 1.33 for polyatomic gas

GLOSSARY

heat Q: energy that flows across a system boundary as the result of a temperature difference *alone*.

work W: any energy flowing across a system boundary that is *not* heat.

first law of thermodynamics: $\Delta U = Q + W$, which follows from the definitions of heat and work and the law of conservation of energy.

heat engine: a device that converts heat energy into work energy, often using a cyclic gas process like those considered in section T3.5.

macroscopic state (of gas): any description of a gas at a certain time that completely specifies its macroscopic properties. In general, for a gas in equilibrium, one would have to specify the type of gas, P, V, N, T, and U. For an ideal gas, however, specifying P, V, N and the type of gas is sufficient: the gas' other properties are determined by these.

PV diagram: a plot of the gas' pressure versus its volume. If N is fixed and known, the macroscopic state of an ideal gas in equilibrium at a given time is represented by a *point* on such a diagram.

quasistatic process: a gas process that occurs so slowly that the gas remains uniform and essentially in equilibrium all the time. As the state of the gas changes during a quasistatic process, it traces a curve on a *PV* diagram.

constrained process: a process that takes a gas through a sequence of states determined by some constraint placed on the gas (such as "P must remain constant").

isochoric process: a constrained gas process where the gas' volume V is held constant.

isobaric process: a constrained gas process where the gas' pressure P is held constant.

isothermal process: a constrained gas process where the gas' temperature is held constant.

adiabatic process: a constrained gas process where heat is not allowed to flow into or out of the gas.

TWO-MINUTE PROBLEMS

T3T.1 In each of the processes described below, energy is being transformed into or from thermal energy. Classify the energy flow in each as being heat (A) or work (B)
  a. Your car's brakes get hot when used repeatedly.
  b. Your pizza gets warm in a microwave oven.
  c. An electric stove element gets hot when turned up.
  d. Your car gets hot in the sun on a relatively cool day.
  e. You get cooler when standing in the breeze from a fan.

T3T.2 A gas with a pressure of 100 kPa is in a container that is a cube 10 cm on a side. If we move one wall in one millimeter, does work flow into (A) or out of (B) the gas? What is the magnitude of W?
  A. 1000 J   D. 1 J
  B. 100 J   E. 0.001 J
  C. 10 J   F. other (specify)

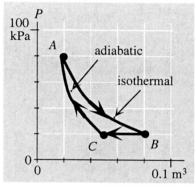

Figure T3.9

T3T.3 The process $B \rightarrow C$ shown in Figure T3.9 is an
  A. isochoric process   D. adiabatic process
  B. isothermal process   E. isometric process
  C. isobaric process   F. none of the above

T3T.4 The signs of Q, W, and ΔU for the process $A \rightarrow B$ are
  A. 0, −, −   D. +, +, 0
  B. 0, +, +   E. −, +, 0
  C. +, −, 0   F. other (specify)

T3T.5 The signs of Q, W, and ΔU for the process $B \rightarrow C$ are
  A. 0, −, −   D. +, +, 0
  B. 0, +, +   E. −, +, 0
  C. +, −, 0   F. other (specify)

T3T.6 The signs of Q, W, and ΔU for the process $C \rightarrow A$ are
  A. 0, −, −   D. +, +, 0
  B. 0, +, +   E. −, +, 0
  C. +, −, 0   F. other (specify)

T3T.7 Is the work energy flowing into or out of the gas in process $B \rightarrow C$ positive (A) or negative (B)? Which of the values below is closest to the magnitude of W?
  A. 0.6 J   C. 300 J   E. 1500 J
  B. 1.5 J   D. 600 J   F. 3000 J

T3T.8 Imagine that a bubble of helium (a monatomic gas) rising from the bottom of the ocean expands in volume by a factor of 8 by the time it reaches the surface (where the pressure is 1 atm). Assume that the bubble rises so fast that is expands essentially adiabatically. What was the pressure on the gas at the depth where it formed (in atm)?
  A. 3.5 atm   C. 16 atm   E. 32 atm
  B. 8 atm   D. 18 atm   F. other (specify)

T3T.9 If the temperature of the bubble described in the previous problem was 320 K when it formed, what is its approximate final temperature when it reaches the surface?
  A. 40 K   C. 320 K   E. 2560 K
  B. 80 K   D. 1280 K   F. other (specify)

HOMEWORK PROBLEMS

BASIC SKILLS

T3B.1 A gas is confined to a cylinder by a piston. The gas has an initial pressure of 120 kPa and a volume of 100 cm³. The piston is slowly moved back until the gas' volume has increased by 0.5%. What is the approximate work that flows into or out of the gas? You can consider this a "sufficiently small" change in volume. (Be sure to give the correct sign as well as the correct magnitude.)

T3B.2 A gas is confined to a cylinder by a piston. The gas has an initial pressure of 95 kPa and a volume of 300 cm³. The piston is slowly pushed in until the gas's volume has decreased by 1%. What is the approximate work that flows into or out of the gas? You can consider this a "sufficiently small" change in volume. (Be sure to give the correct sign as well as the correct magnitude.)

T3B.3 An ideal gas in a cylinder is allowed to expand while its temperature is held fixed. Is there any heat flow involved in this process? If so, does heat energy flow into or out of the gas?

T3B.4 An ideal gas in a cylinder is slowly compressed while its pressure is held fixed. Is there any heat flow involved in this process? If so, does heat energy flow into or out of the gas?

T3B.5 An ideal gas with an initial pressure of 120 kPa is confined to a cylinder with a volume of 150 cm³. We then allow it to slowly expand to a volume of 350 cm³ while adding enough heat to keep its pressure fixed. What is the work that flows into or out of the gas (be sure to give the correct sign as well as magnitude)?

T3B.6 An ideal gas with an initial pressure of 80 kPa is confined to a cylinder with a volume of 150 cm³. We then compress it slowly to a volume of 50 cm³ while keeping its temperature constant. What is the work that flows into or out of the gas (be sure to give the correct sign as well as magnitude)?

T3B.7 A monatomic ideal gas with an initial pressure of 80 kPa is confined to a cylinder with a volume of 600 cm³. We then compress the gas isothermally until its volume has decreased to 450 cm³. What is its pressure now?

T3B.8 A monatomic ideal gas with an initial pressure of 60 kPa is confined to a cylinder with a volume of 600 cm³. We then compress the gas adiabatically until its volume has decreased to 450 cm³. What is its pressure now?

SYNTHETIC

T3S.1 When a substance is heated, does its temperature always change? If not, doesn't this contradict the definition of heat as energy that flows because of a temperature difference? Explain.

T3S.2 When a gas is compressed, does this *always* make work energy flow into the gas? Does this always cause the gas' temperature to rise? Explain.

T3S.3 One mole of helium gas in a cylinder is allowed to expand adiabatically. During the expansion its temperature falls from 310 K to 265 K. How much work energy flows in this expansion (be sure to give both the correct magnitude and the correct sign)?

T3S.4 3.0 × 10²² molecules of nitrogen gas at 280 K is constrained to expand isothermally to 3 times its original volume. Heat must enter the gas during this process. Why? How much heat enters the gas in this process?

T3S.5 A gas is constrained to follow the three-step cyclic process shown in Figure T3.10 below. Prepare a chart (like the one shown in Example T3.1) that specifies the sign of Q, W, and ΔU for each step in the process. What is the net work flowing into or out of the gas for the entire cyclic process (be sure to give the correct magnitude and sign)?

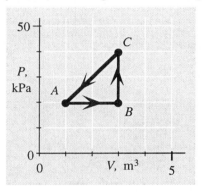

Figure T3.10

T3S.6 A gas is constrained to follow the three-step cyclic process shown in Figure T3.11 below. Prepare a chart (like the one shown in Example T3.1) that specifies the sign of Q, W, and ΔU for each step in the process.

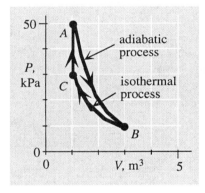

Figure T3.11

T3S.7 Heat must flow into the gas during the process $C \rightarrow A$ in Figure T3.11. Why? If the gas' temperature is 290 K at point C, find its temperature at point A and the heat that has flowed into the gas in this process.

T3S.8 A bubble of air is formed at the bottom of the ocean floor 66 ft below the surface, where the ambient pressure is about 300 kPa = 3 atm. The bubble has an initial volume of about 25 cm³ and a temperature of 8°C. If it rises so fast that it expands essentially adiabatically, what is its final volume? What is its final temperature?

T3S.9 A research balloon bound for the stratosphere is filled at sea level with 800 m³ of helium whose initial temperature is 285 K. The balloon is released, and climbs to an altitude where the air pressure is 0.045 times its sea-level value. If the helium expands adiabatically, what is the balloon's volume now? What is the helium's temperature?

T3S.10 Use equation T3.18 to show that the work done during an adiabatic volume change from V_i to V_f is

$$W = \frac{P_i V_i^{\gamma}}{\gamma - 1} \left(\frac{1}{V_f^{\gamma - 1}} - \frac{1}{V_i^{\gamma - 1}} \right) \qquad \text{(T3.20)}$$

(*Hint*: Solve T3.18 for P as a function of V and then use equation T3.8.)

RICH-CONTEXT

T3R.1 Imagine that the atmospheric pressure at the top of a tall mountain is about 0.65 times the pressure at sea level. If it is 30°C (86°F) at the beach, and a stiff breeze blows this air up the mountain so rapidly that the air essentially expands adiabatically, what is the approximate temperature at the top of the mountain?

T3R.2 Imagine that we have a sample of nitrogen gas confined to a cylinder by a moveable piston. The cylinder is immersed in ice water and the gas initially has a temperature of 0°C. You (1) fairly rapidly compress the gas, adiabatically decreasing its volume by a factor of two. You then

(2) hold the piston still until the gas has cooled again to 0°C, and then (3) allow the gas to expand slowly to its original volume while allowing plenty of time for heat to move out of or into the gas (so that its temperature remains very close to 0°C). If 85 g of ice was melted during the second step of the cycle, how much work energy did you put into the gas during the first step? Describe your reasoning. (*Hint*: I suggest that you first draw a *PV* diagram. Also, it only *looks* like you do not have enough information to solve this problem. If you find yourself doing lots of calculations, stop and *think* about it some more.)

ADVANCED

T3A.1 One can find the work involved in an adiabatic process in one of either two ways. The first way is to use equation T3.20. The second way is to realize that since no heat flows into the gas, the work that flows into (or out of the gas) is the same as the gas' change in internal energy: $\Delta U = W$. You can find ΔU using equation T3.19 and the results of the last chapter. Prove mathematically that this method yields the same results as equation T3.20.

ANSWERS TO EXERCISES

T3X.1 *a*: H, *b*: H, *c*: W, *d*: H, *e*: W, *f*: H, *g*: H, *h*: W.

T3X.2 When we compress a gas, its pressure will generally change, but equation T3.7 assumes that the pressure is fixed during the volume change dV. Therefore T3.7 only works if the change in volume is sufficiently small that the pressure does not change significantly during the process.

T3X.3 Substituting as suggested we get:

$$W = -\int_{V_i}^{V_f} \frac{N k_B T}{V} dV = -N k_B T \int_{V_i}^{V_f} \frac{dV}{V}$$

$$= -N k_B T [\ln V_f - \ln V_i] = -N k_B T \ln(V_f / V_i) \quad \text{(T3.21)}$$

T3X.4 $W = +69$ J.

T3X.5 The chart should look like this:

Step:	Q	W	ΔU
$A \rightarrow B$	+	−	+
$B \rightarrow C$	−	0	−
$C \rightarrow A$	0	+	+

Net work done in cycle:

The area inside the cycle in the diagram is about 5.5 squares. Each square corresponds to

$(2 \times 10^4 \text{ N/m}^2)(0.05 \text{ m}^3)$
$= 1000 \text{ N·m} = 1000 \text{ J}$

so $W \approx 5500$ J.

T3X.6 The computed value is 2200 J, about 8% smaller.

T3X.7 Since $PV = N k_B T$, we have

$$P = \frac{N k_B T}{V} \qquad \text{(T3.22)}$$

If T were to remain constant, then P would be proportional to $1/V$, but if T also falls as the gas expands, then P will be smaller at any given V than it would be if T remained fixed.

T3X.8 Plugging in as directed, we get $\frac{4}{2} N k_B dT = -P\, dV$. Multiplying both sides by $2/n$ yields equation T3.16.

T3X.9 Plugging $N k_B\, dT = -(2/n)P\, dV$ into equation T3.14, we get $P\, dV + V\, dP = -(2/n)P\, dV$. Rearranging terms, we get

$$V\, dP = -(1+2/n)P\, dV \qquad \text{(T3.23)}$$

Dividing both sides by PV yields the desired result.

MACROSTATES AND MICROSTATES

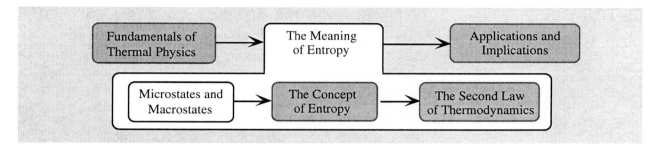

T4.1 OVERVIEW

In the past three chapters, we have learned much about thermal energy, how it is stored at the microscopic level, how it can be transferred from one object to another, and how it is linked to an object's temperature. This helps us understand *some* things about the "paradigmatic thermal process" first described in Chapter T1. We know, for example, that it is *energy* that flows from the hot object to the cold object, and we know why the hot object's temperature decreases (and the cold object's temperature increases) as this energy flows. We can perhaps imagine how vigorously wiggling atoms in the hot object bumping into atoms in the cold object might transfer energy to that object. Even so, we still cannot answer the basic question of this unit: *why does the energy spontaneously flow from the hot object to the cold object and never the other way around?*

It turns out that there is a profound and (with hindsight) very *simple* reason for this irreversible behavior, and it will be our task in the next three chapters to understand this reason. In this chapter, we will mostly lay foundations, defining the basic concepts of *macrostate*, *microstate* and *macropartition* and developing a simplified model for macroscopic objects that will help us understand the entropy concept. In Chapter T5, we will use a computer program to help us explore what happens when two objects are brought into thermal contact: this will lead us to the definition of a new concept called entropy. In Chapter T6, we will continue to explore the implications of the entropy concept, ultimately summarizing our findings in the Second Law of Thermodynamics.

T4.2 *HOW MANY MICROSTATES IN A MACROSTATE?* introduces the crucial concepts of *microstate*, *macrostate* and *multiplicity*.

T4.3 *THE EINSTEIN MODEL OF A SOLID* presents an idealized model for a solid that is simple enough that we can analyze its behavior fully but complex enough to qualitatively represent the behavior of real objects.

T4.4 *COUNTING MICROSTATES* illustrates how we can (in principle) count the microstates of our idealized model.

T4.5 *TWO EINSTEIN SOLIDS IN THERMAL CONTACT* explores what happens when we put two model solids in thermal contact.

T4.6 *THE FUNDAMENTAL ASSUMPTION* states the basic assumption at the heart of statistical mechanics and begins to explore its implications.

T4.2   HOW MANY MICROSTATES IN A MACROSTATE?

Definition of *macrostate*

In the previous chapter, we saw that we could specify the state of a sample of a given type of ideal gas by specifying any three of the four quantities P, V, N, and T (the fourth quantity can always be computed from the others using the ideal gas law $PV = Nk_BT$). Two samples of a given type of ideal gas having the same values of P, V, N and T are thermodynamically indistinguishable. On the other hand, we can easily tell the difference between two otherwise identical bottles of gas having different values of N, T, P or V without having to examine the gases at the microscopic level.

The gas state specified by P, V, N, and T is a specific example of what we will call a **macrostate**. We say that two systems have the same macrostate if they cannot be physically distinguished by any macroscopic experiment (that is, any experiment that does not involve looking at the behavior of individual molecules in the system). The macrostate of a given system is usually specified by stating values for a certain set of macroscopically measurable quantities, (P, V, N, and T in the case of an ideal gas).

Different sets of variables might be appropriate for different kinds of systems. As a second example, imagine a solid object of a specified material. The volume V of a solid material is usually directly proportional to N and depends only weakly on T and P. Indeed, a solid's internal pressure P typically is the same as atmospheric pressure (since most solids we work with will be directly exposed to the atmosphere) which in turn is essentially fixed. So to a good degree of accuracy, we can specify the macrostate of a solid of a given material under ordinary conditions by specifying just N and T.

Definition of *microstate*

The **microstate** of a system, on the other hand, is specified by describing in detail the state of each molecule in the system. In classical mechanics, the state of a particle is described by six numbers: its position coordinates x, y, z, and its momentum components p_x, p_y, p_z (given these numbers at any time, we can predict the future motion of the particle using Newton's laws). So to specify the classical microstate of a gas or solid containing N molecules, we have to specify a total of $6N$ numbers. Note that while this idea is *conceptually* straightforward, there are so many molecules in even the tiniest speck of substance that it would be impossible in *practice* to measure these numbers at a given time. (Even simply writing them down would require more time than the projected lifetime of the sun.)

Actually, we should describe the physical state of an object as small as a molecule using *quantum* mechanics, not classical mechanics. In quantum mechanics, the state of a given quanton in a gas or solid is described by specifying its energy level. To a good degree of approximation, we can treat a gas molecule in a bottle as if it were a "quanton in a box" (in the language of Unit Q), where the box in this case is the container holding the gas. The energy state of a quanton moving in a one dimensional box is specified by a single integer n that states how many half-wavelengths of the quanton's wavefunction fit between the box boundaries. For a quanton moving in a 3-dimensional box, it turns out that we need to provide three integers n_x, n_y, and n_z (that describe the number of half-wavelengths of the quanton's wavefunction that fit between the walls of the box along the x, y, and z directions respectively) to completely specify its quantum state. So to describe the microstate of the gas, we need to specify these three numbers for every quanton in the gas. This is a total of $3N$ numbers, which is only marginally less daunting than describing the classical microstate.

There are *many* microstates in a macrostate

The most important thing to understand about microstates and macrostates is that a gas (or anything else) in a given, well-specified macrostate could be in any one of a *huge* number of different microstates that we are unable to distinguish by macroscopic measurements. For example, consider specifying the macrostate of an ideal monatomic gas by stating values for T, V, and N. Since the thermal energy U of an ideal gas depends only on T and N ($U = \frac{3}{2} N k_B T$ for an ideal monatomic gas), specifying T is essentially the same as specifying U. Knowing the gas' *total* thermal energy, however, does not tell us anything about how that energy is distributed among its molecules. For a given value of U, there are a myriad of possible microstates (each corresponding to a different way of distributing the energy among the molecules) that nonetheless add up to the same total U. Moreover, even if we fix how much *energy* each molecule has, we still do not know the direction of its motion. The fact that we can arbitrarily choose the *signs* of each of a molecule's three momentum components without changing its energy *alone* contributes a factor of 2^{3N} to the number of possible microstates consistent with the single macrostate specified by values of U and N.

The point is that for any complex system, there are an immense number of possible microstates consistent with any given macrostate. The answer to the question posed by the title of this section is (at this point), "there are many, many, many microstates in a macrostate."

Exercise T4X.1: Imagine that we have a gas with a known total energy U, consisting of only $N = 10$ molecules. What is the value of the factor 2^{3N} associated with our freedom to choose the signs of momentum components? Would you consider this a large number?

T4.3 THE EINSTEIN MODEL OF A SOLID

To get more specific about how many microstates there are in a macrostate, we need to talk about a specific kind of system. For the remainder of this chapter (and the next), I want to focus on a particular simplified model for a solid.

In 1907, Albert Einstein published a paper that proposed a simple but reasonably accurate model for predicting the thermal behavior of elemental crystalline solids (such as crystals of pure aluminum, copper, carbon, or gold). Einstein proposed that we treat the atoms in such a solid as if they were held in their lattice position by springs, as shown in Figure T4.1 (this is essentially the *bedspring model* introduced in chapter T2). In a real solid, the atoms are actually held in place by interatomic electrical interactions, but for small amplitude oscillations, the potential energy functions for such interactions become approximately the same as that for a simple mass on a spring (as discussed in unit C).

Einstein model: atoms in crystal behave like 3D harmonic oscillators

Each atom in this model is free to oscillate in three dimensions. In both newtonian and quantum mechanics, however, we can treat a three-dimensional oscillating object as if it were oscillating *independently* in one dimension along each of the three coordinate axes. For example, in newtonian mechanics, the total energy of a three-dimensional harmonic oscillator would be written

We can treat a 3D oscillator as three 1D oscillators

$$E = \tfrac{1}{2} mv^2 + \tfrac{1}{2} k_s r^2 = \tfrac{1}{2} m(v_x^2 + v_y^2 + v_z^2) + \tfrac{1}{2} k_s (x^2 + y^2 + z^2)$$

$$= (\tfrac{1}{2} mv_x^2 + \tfrac{1}{2} k_s x^2) + (\tfrac{1}{2} mv_y^2 + \tfrac{1}{2} k_s y^2) + (\tfrac{1}{2} mv_z^2 + \tfrac{1}{2} k_s z^2) \qquad \text{(T4.1)}$$

where r is the distance that the oscillating object (in this case the atom) is from its equilibrium position. Note how the terms in the energy can be grouped into three pairs, each pair of which would be the energy associated with a *one* dimensional oscillation along one of the coordinate axes, without any reference to what is happening in the other coordinate directions. You can also show from this equation (if you want: see Problem T4A.1) that the atom's motion along each coordinate axis is exactly as if the atom were oscillating in one dimension along that axis alone. (This is a special property of the harmonic oscillator potential energy function: most other potential energy functions cannot be pulled apart this way.)

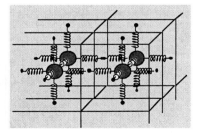

Figure T4.1: Einstein's model of a crystalline solid. In this model, atoms are imagined to be held in their position by springs.

According to quantum mechanics, the energy associated with each of these separate one-dimensional oscillations is quantized, so that the atom's total vibrational energy is given by

$$E = \hbar\omega(n_x + \tfrac{1}{2}) + \hbar\omega(n_y + \tfrac{1}{2}) + \hbar\omega(n_x + \tfrac{1}{2}) \tag{T4.2a}$$

where $\omega = [k_s / m]^{1/2}$, k_s being the effective spring constant of the interactions holding the atom in its place and m being the atom's mass (we are assuming that k_s and thus ω are the same for all three coordinate directions here). The quantities n_x, n_y, and n_z here are independent, non-negative integers (0, 1, 2, 3, and so on). Each of the three terms here is the same as the expression for the energy of a *one*-dimensional quantum harmonic oscillator (as we discussed in Unit Q).

We can rewrite this equation as follows

$$E = \sum_{i=1}^{3} \hbar\omega(n_i + \tfrac{1}{2}) = \sum_{i=1}^{3} \varepsilon(n_i + \tfrac{1}{2}) \tag{T4.2b}$$

where $\varepsilon \equiv \hbar\omega$ is the energy difference between adjacent levels of each one-dimensional oscillation and n_1, n_2, n_3 just are a different way of labeling the integers n_x, n_y, and n_z. We can then find the solid's total thermal energy by summing over all atoms (three terms per atom)

$$U = \sum_{i=1}^{3N} \varepsilon(n_i + \tfrac{1}{2}) = \sum_{i=1}^{3N} \varepsilon n_i + \sum_{i=1}^{3N} \tfrac{1}{2}\varepsilon = \sum_{i=1}^{3N} \varepsilon n_i + \frac{3N}{2}\varepsilon \tag{T4.3}$$

The quantum zero energy of the solid is irrelevant

Now, the $3N\varepsilon/2$ term in this equation is called the **zero-point energy** of the solid. The solid will have this energy at absolute zero (which is the temperature where all atoms are in their *lowest possible* energy state) and cannot be affected by *any* thermal process. This means that as far as thermal processes are concerned, the *effective thermal energy* of the solid (that is, the part of its internal energy that changes with temperature) is

The quantum thermal energy of an Einstein solid

$$U = \sum_{i=1}^{3N} \varepsilon n_i \tag{T4.4}$$

(Another way to justify removing this term is that the energy of an object is only defined up to a constant anyway, as we discussed in unit C. We can thus take advantage of our freedom to choose this overall constant to *define* the thermal energy U to be zero at absolute zero. Either way we look at it, the constant $3N\varepsilon/2$ is irrelevant for the thermal physics of the solid.)

The basic point of equation T4.4 is that *a crystalline solid containing N atoms behaves as if it contained 3N identical independent quantum harmonic oscillators, each of which can store an integer number n_i of energy units ε.* This is the model proposed by Einstein in his 1907 paper. In his honor, we will call a solid that behaves in a manner consistent with this model an **Einstein solid**.

Experimental support

Table T4.1: Values of dU/dT for various elemental solids.

Solid	dU/dT per mole
Lead	26.4 J/K·mole
Gold	25.4 J/K·mole
Silver	25.4 J/K·mole
Copper	24.5 J/K·mole
Iron	25.0 J/K·mole
Aluminum	26.4 J/K·mole

(Adapted from Serway, *Physics*, 3/e, p. 530.)

One of the practical consequences of this model is that when the temperature is high enough so that the quantum oscillators are completely unfrozen, each of the $3N$ one-dimensional oscillators in the solid has (according to the last chapter) two associated degrees of freedom, each of which stores $\tfrac{1}{2}k_BT$ of energy on the average. This means that we would expect that at sufficiently high temperatures

$$U = (3N)(2)\tfrac{1}{2}k_BT = 3Nk_BT \quad\Rightarrow\quad \frac{dU}{dT} = 3Nk_B \tag{T4.5}$$

For a solid containing one mole of atoms, $3N_Ak_B = 24.9$ J/K. When we measure dU/dT per mole for various types of elemental solids, we find reasonably close agreement with this result (see Table T4.1), implying that the model is basically sound for this type of solid.

T4.4 COUNTING MICROSTATES

From this section to the end of the next chapter, we will focus our entire attention on the thermal behavior of this particular model for a solid. Why focus on this model, and not (for example) the ideal gas model? The answer is that it is *much* easier to determine the number of microstates corresponding to each macrostate for an Einstein solid than for any other reasonably realistic model. This in turn makes it comparatively easier to determine what statistical physics predicts about this model. Moreover, all other models of complex systems are *qualitatively* similar to this model, so what we learn from close inspection of this model applies in qualitative terms at least to other systems as well (as we will discuss in chapter T6).

Why focus on Einstein solids?

To describe the macrostate of an Einstein solid, it is sufficient to specify the number of atoms N and its total internal energy U. Since $U = 3Nk_BT$ for an Einstein solid, if we know U, we know T. In this simplified model, each atom occupies a fixed volume and is independent of other atoms, so V depends on N but not on temperature T or pressure P or any other factors. (The volume of a typical *real* solid depends only very weakly on temperature and pressure, so our model is a reasonably good approximation.)

Describing the macrostate of an Einstein solid

To describe a microstate of the solid, we have to specify an integer value n_i for each of the solid's $3N$ independent oscillators. There will in general be *many* microstates (that is many distinct sets of values for all $3N$ integers n_i) that add up in equation T4.4 to the same total U. The number of possible microstates corresponding to the same given macrostate is called the **multiplicity** g of that macrostate. In the case of an Einstein solid, where the macrostate is specified by U and N:

Definition of *multiplicity*

$g(N,U) \equiv$ multiplicity of the macrostate specified by U and N

$\quad\quad$ = number of N-atom microstates having total energy U $\quad\quad$ (T4.6)

How can we determine $g(U,N)$ for given values of U and N? The beauty of the Einstein solid model is that this is not *conceptually* difficult. According to equation T4.4, the total energy in an Einstein solid is an integer multiple of the basic energy unit ε. Think of each energy unit as a marble, and each of the $3N$ oscillators in the solid as a bin in which we can put an integer number of marbles. When we specify the total energy of the solid, we are essentially specifying the total number $n \equiv U/\varepsilon$ of marbles that we have to spread around. Counting the microstates for this U, then, is the same as counting how many different ways we can divide n marbles into the $3N$ bins that can hold these marbles.

Counting microstates

In a solid of macroscopic size, $3N$ will be on the order of magnitude of 10^{23}, but let's start small. Imagine we have an Einstein solid consisting of a *single* atom, that is, *three* independent oscillators. (Of course, a single atom will have no lattice in which it can oscillate, but let's just pretend that this makes sense. We'll work up to larger numbers of atoms shortly.) We can describe the microstate of this system by specifying energy level integers for each of the three oscillators (that is, how many "energy marbles" each of these three bins contains). Let's write these numbers as a triplet of digits: for example, the triplet 032 specifies the microstate where the first oscillator has zero units of energy, the second has three units, and the third has two units. The total energy contained in the system in this case is $U = 5\varepsilon$. Other possible microstates corresponding to this macrostate are 320, 230, 302, 203, 023, 113, 311, 131, 041, 014, etc.

Examples of counting microstates for small *U, N*

So let's start counting microstates for various different macrostates of this hypothetical one-atom Einstein solid. First of all, imagine that the total energy of the solid has its lowest possible value $U = 0$. There is only one microstate (000) compatible with this total energy, so we say that the multiplicity of this macrostate is $g(N, U) = g(1, 0\varepsilon) = 1$.

Now imagine that the total energy of the solid is $U = \varepsilon$ (that is, the solid contains exactly one unit of energy). The microstates compatible with this total energy are 100, 010, and 001, a total of three. The multiplicity of this macrostate is thus $g(1, 1\varepsilon) = 3$.

Exercise T4X.2: Now imagine that the total energy of the solid is $U = 2\varepsilon$. The multiplicity of this macrostate turns out to be $g(1, 2\varepsilon) = 6$. Write down the triplets for the six microstates corresponding to this macrostate.

Exercise T4X.3: Now imagine that the total energy of the solid is $U = 3\varepsilon$. What is the multiplicity of this macrostate? (Write down all possible microstate triplets consistent with this macrostate and count them up.)

Similarly, $g(1, 4\varepsilon) = 15$, $g(1, 5\varepsilon) = 21$, and $g(1, 6\varepsilon) = 28$.

Similarly an Einstein solid consisting of two atoms (6 independent oscillators) and zero total energy has one microstate 000000, so $g(2, 0\varepsilon) = 1$. If it has a total energy of $U = \varepsilon$, then the six possible microstates are 000001, 000010, 000100, 001000, 010000, and 100000, meaning that $g(2, 1\varepsilon) = 6$. If it has a total energy of $U = 2\varepsilon$, then the possible microstates are 000002, 000020, 000200, 002000, 020000, and 200000, 000011, 000101, 001001, 010001, 100001, 000110, 001010, 001100, 010010, 010100, 011000, 100010, 100100, 101000, and 110000, for at total of 21 states, so $g(2, 2\varepsilon) = 21$. In a similar fashion, we can determine that $g(2, 3\varepsilon) = 56$, $g(2, 4\varepsilon) = 126$, and so on.

You can see that this counting of microstates gets pretty tedious after a while. There is in fact a general formula for the number of microstates for a given system. If M is the number of oscillators in the solid (not atoms: $M = 3N$), and n is the total number of energy units ε to be distributed among these oscillators ($U = n\varepsilon$), then the multiplicity of the macrostate is:

The multiplicity of a Einstein solid macrostate

$$g(N, U) = \frac{(M + n - 1)!}{n!(M - 1)!} = \frac{(3N + U/\varepsilon - 1)!}{(U/\varepsilon)!(3N - 1)!} \qquad \text{(T4.7)}$$

where $n! \equiv n$ **factorial** $= 1 \cdot 2 \cdot 3 \cdot \ldots \cdot (n-1) \cdot n$. (The derivation of this formula is discussed in problem T4A.2.) So, for example,

$$g(2, 4\varepsilon) = \frac{(6 + 4 - 1)!}{4!(6 - 1)!} = \frac{9!}{4!5!} = \frac{9 \cdot 8 \cdot 7 \cdot 6 \cdot 5 \cdot 4 \cdot 3 \cdot 2 \cdot 1}{(4 \cdot 3 \cdot 2 \cdot 1)(5 \cdot 4 \cdot 3 \cdot 2 \cdot 1)}$$

$$= \frac{9 \cdot 8 \cdot 7 \cdot 6}{4 \cdot 3 \cdot 2 \cdot 1} = 9 \cdot 7 \cdot 2 = 126 \qquad \text{(T4.8)}$$

Exercise T4X.4: Check that this formula yields the same results for $g(1, 0)$, $g(1, \varepsilon)$, $g(1, 2\varepsilon)$ and $g(2, 3\varepsilon)$ that we found earlier by direct counting. Verify that $g(1, 6\varepsilon) = 28$. Also, what is the multiplicity of an Einstein solid with three atoms and eight units of energy?

T4.5 TWO EINSTEIN SOLIDS IN THERMAL CONTACT

Suppose that we now bring two Einstein solids A and B, one with N_A atoms ($3N_A$ oscillators) and one with N_B atoms ($3N_B$ oscillators), into close thermal contact, so that energy can flow freely back and forth between them. The ultimate question that we seek to resolve is how these solids will behave macroscopically after they have been brought into contact.

The macrostate of solid A is specified by N_A and U_A, while the macrostate of solid B is specified by N_B and U_B. Since N_A and N_B are fixed, the macrostates of A and B are essentially determined by their respective energies U_A and U_B. If the combined system of the two solids is thermally isolated, its total energy $U = U_A + U_B$ is fixed (by conservation of energy), but at least in principle, the energies U_A and U_B of the two solids could have any values consistent with that total. For example, if the combined system's total energy is $U = 6\varepsilon$, then possible pairs of values for U_A and U_B include $U_A = 0$ and $U_B = 6\varepsilon$, or $U_A = 2\varepsilon$ and $U_B = 4\varepsilon$ or $U_A = 5\varepsilon$ and $U_B = \varepsilon$, and so on.

Let us call a given pair of macrostates for solids A and B that are consistent with $U = U_A + U_B$ having a given fixed value a **macropartition** of the combined system for that U. For example, the pair of macrostates where $U_A = 2\varepsilon$ and $U_B = 4\varepsilon$ is one possible *macropartition* of the combined system for $U = 6\varepsilon$.

Macropartition **of a pair of objects in thermal contact**

Different macropartitions of the combined system of two solids therefore amount to different ways that the energy can be *macroscopically* divided between the solids. There is a real distinction to be made here between a *macropartition* and a *microstate* of the combined system. A microstate of the combined system specifies exactly how much energy *each individual oscillator* in both solids has. A macropartition, on the other hand, only specifies the macroscopic total energy U_A and U_B that each of the two macroscopic solids have. In the latter case, we do not need to use a microscope to peer at the individual atoms: we simply measure the temperature of each solid (higher temperatures reflect higher internal energies), which is a *macroscopic* measurement.

It is easiest to illustrate these ideas with a specific example. Suppose we bring two hypothetical one-atom solids into thermal contact, and imagine that their total combined energy is $U = 6\varepsilon$. The table below lists all possible macropartitions of the combined system with this total energy:

A macropartition table for Einstein solids in contact

Table T4.2: Possible macropartitions for $N_A = N_B = 1$, $U = 6\varepsilon$.

Macropartition	U_A	U_B	g_A	g_B	g_{AB}
0:6	0	6	1	28	28
1:5	1	5	3	21	63
2:4	2	4	6	15	90
3:3	3	3	10	10	100
4:2	4	2	15	6	90
5:1	5	1	21	3	63
6:0	6	0	28	1	28

Grand total number of microstates = 462

The numbers under the columns labeled U_A and U_B specify the macrostates of solids A and B by specifying the total energy of each solid in units of the basic energy unit ε. Each number under the heading g_A states the multiplicity of solid A when it has the specified energy U_A, and each number under the heading g_B states the same for solid B. The value of g_{AB} specified for each macropartition is the total multiplicity of the combined system in that macropartition (that is, the total number of distinct microstates available to the system as a whole in that macropartition). This total multiplicity g_{AB} is the product of g_A and g_B:

$$g_{AB} = g_A \cdot g_B \qquad \text{(T4.9)}$$

because for each one of the g_A possible microstates for solid A, there are g_B possible microstates for solid B. For example, in the macropartition 3:3, the possible microstates of solid A are (in our previous notation) 300, 030, 003, 210, 201, 021, 120, 102, 012 and 111, and the possible microstates of system B are the same. The possible microstates of the combined system are as follows (the triplets on the left and right specify the microstates of solids A and B respectively): 300-300, 300-030, 300-003, 300-210, 300-201, 300-021, 300-120, 300-102, 300-012, 300-111, 030-300, 030-030, 030-003, 030-210, and so on, for a total of $10 \times 10 = 100$ distinct microstates.

Exercise T4X.5: Prepare an analogous table for the case where $N_A = N_B = 1$ and $U = 8\varepsilon$. (Most of the multiplicities g_A and g_B you can copy from Table T4.2: use equation T4.7 to calculate the rest).

T4.6 THE FUNDAMENTAL ASSUMPTION

In a more familiar system such as the ideal gas, it is obvious that the gas will continually and randomly shift from microstate to microstate as the molecules collide with each other and exchange energy. A similar situation will apply here if we assume that the elementary oscillators in each solid weakly interact with each other: not so strongly that the approximation that the oscillators are independent becomes invalid, but strongly enough so that energy is freely interchanged between the oscillators. Energy will also be shifted randomly back and forth across the boundary between the solids through the interactions of the oscillators on the surfaces in contact.

In short, as time passes, the combined system of two solids will randomly shift between different microstates consistent with the constraint that the total energy have some fixed value U. This means that under some circumstances, the macropartition of the combined system might fluctuate as the system randomly samples microstates in different macropartitions. For example, in the situation considered in Table T4.2, the combined system in microstate 012-300 (one of the microstates corresponding to macropartition 3:3), might evolve to 013-200 (one of the microstates corresponding to macropartition 4:2) by transfer of one unit of energy across the boundary. In time, this system will sample each of the 462 possible microstates, and thus each of the possible macropartitions.

Fundamental assumption of statistical mechanics

Now comes the big question: Can we say something about which of the macropartitions are the ones that we are most likely to see if we were to peek at the system at various times? The answer is yes, if we are willing to accept a simple and plausible assumption about the behavior of such systems. This assumption is called the **fundamental assumption** of statistical mechanics:

> **The fundamental assumption:** A system will shift between its accessible microstates in a totally random fashion, without preferring any one microstate over another. Thus, a system is equally likely to be found in any one of its accessible microstates.

Accessible means in this context "consistent with the value of the total internal energy of the system in question."

This disarmingly simple postulate provides the foundation for understanding irreversible processes, as we will see in the next two chapters. It should be noted that even though this assumption is simple and plausible, its ultimate justification is that it can be used to correctly predict the behavior of macroscopic systems.

The most important consequence of this principle for us is that the probability of occurrence of a given energy macropartition (consistent with the given total internal energy) is directly proportional to the number of microstates that indistinguishably generate that macropartition, that is, to the total multiplicity g_{AB} of that macropartition.

For example, suppose that we were to take a large number of "snapshots" of the system of two Einstein solids described in Table T4.2. The fundamental assumption means that we would find the system to be in 3:3 in about $100/462 = 0.216 = 21.6\%$ of the pictures, macropartition 0:6 (or macropartition 6:0) in about $28/462 = 0.061 = 6.1\%$ of the pictures and so on. Note that macropartition 3:3, the macropartition for which the energy is shared equally between the two identical solids, is the single most probable macropartition of the system.

Exercise T4X.6: Compute the probabilities of each of the macropartitions in Table T4.2, and write them to the right of the value of g_{AB} for that macropartition. Do the same for the table that you constructed when you did Exercise T4X.5. Which macropartition in the latter case is the most probable?

Doing the calculations required to set up a table such as Table T4.2 can be very tedious, particularly as the number of atoms in each solid becomes large. During your class session on this chapter, your instructor should show you how to use a computer program called *StatMech* that does all of the calculations for you. In principle, you can easily do by hand what *StatMech* does (and in fact you did in Exercise T4X.6), but once you understand the basic principle of the calculation, you will appreciate the ease and rapidity with which the computer generates the results. (See section T5.2 for a description of the program.)

In the next chapter, we will use this computer program to explore the consequences of the fundamental assumption in the context of the interacting Einstein solids. Our explorations will lead us naturally to the concept of entropy as a useful way to describe the implications of the fundamental assumption. This, in turn, will help us understand why some processes are irreversible.

I. MACROSTATES AND MICROSTATES   **SUMMARY**
    A. The *macrostate* of a system
        1. This is specified by a sufficient set of macroscopic variables
        2. For example, any three of P, V, N, or T for an ideal gas
        3. Other systems may be described by different variables
    B. The *microstate* of a system
        1. This is specified by describing the quantum state of every molecule
        2. For an ideal gas, this means $3N$ quantum numbers
    C. There are many microstates in a macrostate!

II. EINSTEIN SOLID
    A. An Einstein solid is a simplified model of a solid
        1. It imagines atoms in a crystal to be bound to the lattice by springs
        2. Thus the atoms behave like identical 3D harmonic oscillators
            a. We can treat each 3D oscillator as three 1D oscillators
            b. Each 1D oscillator stores an integer number of energy units
        3. An Einstein solid's total internal energy is thus $U = \sum \varepsilon n_i$ (T4.4)
            a. The sum is from $i = 1$ to $3N$ (that is, over all 1D oscillators)
            b. This formula ignores the zero-point energy (which is fixed)
        4. The equipartition theorem implies that $U = 3Nk_BT$ (T4.5)
           (which seems reasonably consistent with experimental data)
    B. We will focus on Einstein solid because:
        1. It is easy to count microstates of an Einstein solid
        2. The model is *qualitatively* similar to more realistic models
    C. The macrostate of Einstein solid is adequately specified by U and N
    D. The multiplicity $g(N, U)$ of a macrostate of this solid
        1. Specifies number of microstates consistent with this macrostate
        2. $g(N, U = (3N + U/\varepsilon - 1)! \, [(3N-1)! \, (U/\varepsilon)!]^{-1}$ in this case   (T4.7)

III. EINSTEIN SOLIDS IN THERMAL CONTACT
    A. A *macropartition* of a system of two objects in thermal contact
        1. In general: we must specify the macrostate of each object
        2. For two Einstein solids: specifying U and N for each is sufficient
    B. A *macropartition table*: see Table T4.2 (Note: $g_{AB} \equiv g_A \cdot g_B$)
    C. The *fundamental assumption of statistical mechanics*
        1. Definition: each accessible microstate is equally probable
        2. $\Rightarrow$ macropartitions with largest multiplicities are most probable

GLOSSARY

macrostate: the state of a system as described by its macroscopic variables. Two systems have the same macrostate if all macroscopically measurable quantities (such as P, V, N, T, U, and so on) have the same values

microstate: the state of a system specified by describing the quantum state of each molecule in the system

Einstein solid: a microscopic model for a crystalline solid that treats each atom as if it were held in its lattice position by springs (that is, the atoms behave like independent identical 3D harmonic oscillators).

zero-point energy: the energy that a quantum system has when it is in its lowest possible quantum energy state. An Einstein solid has a zero-energy of $3N\hbar\omega/2$, which, since it cannot be changed and does not depend on temperature, is irrelevant to its thermal behavior.

multiplicity g: the number of microstates consistent with a given macrostate.

macropartition: a *pair* of macrostates for two objects in thermal contact consistent with the overall macrostate for the combined system. In the case of a pair of Einstein solids, specifying the macropartition amounts to specifying the energies U_A and U_B of each solid, subject to the constraint that $U = U_A + U_B$ be a fixed constant.

The fundamental assumption (of statistical mechanics): asserts that each accessible microstate of a system is equally probable. *Accessible* here means that the microstate must be consistent with the known macrostate of the system. For a combined system of two objects in thermal contact, this means that the probability that a given macropartition will be observed is proportional to its multiplicity.

TWO-MINUTE PROBLEMS

T4T.1 Consider a system consisting of two Einstein solids P and Q in thermal contact. Assume that we know the number of atoms in each solid and ε. What do we know if we also know the quantum state of each atom in each solid?
A. the system's macrostate
B. the system's microstate
C. the system's macropartition
D. B and C
E. A and C
F. $A, B,$ and C

T4T.2 Consider a system consisting of two Einstein solids P and Q in thermal contact. Assume that we know the number of atoms in each solid and ε. What do we know if we also know the total energy in each of the two objects?
A. the system's macrostate
B. the system's microstate
C. the system's macropartition
D. B and C
E. A and B
F. $A, B,$ and C

T4T.3 Consider a system consisting of two Einstein solids P and Q in thermal contact. Assume that we know the number of atoms in each solid and ε. What do we know if we also know the total energy of the combined system?
A. the system's macrostate
B. the system's microstate
C. the system's macropartition
D. B and C
E. A and C
F. $A, B,$ and C

T4T.4 What is the *crucial* characteristic of an Einstein solid that makes it easier to analyze in the context of this chapter than most other kinds of thermodynamic systems?
A. the atoms are arranged in a regular, cubical lattice
B. the atoms are identical

C. its microstates are comparatively easy to count
D. the atomic vibrational energy levels are equally spaced
E. $g(U,N)$ is always reasonably small
F. other (specify)

T4T.5 The zero-point energy of an Einstein solid can be ignored because
A. it is zero
B. it never changes in any thermal interaction
C. it is insignificant compared to the solid's total energy
D. it is just a quantum-mechanical effect
E. other (specify)

T4T.6 The multiplicity of an Einstein solid with 3 atoms and 4 units of energy is:
A. 4.8×10^8
B. 715
C. 495
D. 36
E. 12
F. other

T4T.7 Which of the following statements is true?
A. there are always many microstates in a macrostate.
B. all accessible macrostates are equally probable.
C. all microstates of a system are equally probable.
D. all accessible macropartitions are equally probable.
E. when two objects in thermal contact are isolated from everything else, their macrostates cannot change
F. none of the above

HOMEWORK PROBLEMS

BASIC SKILLS

T4B.1 Imagine that we have an ideal gas consisting of 15 molecules. We can flip the signs of each of the three momentum components of a given molecule without changing its overall energy (and thus without changing the gas' macrostate). How many possible patterns of sign choices are there? How many times larger is this than the answer to Exercise T4X.1?

T4B.2 Calculate the multiplicity of an Einstein solid with $N = 1$ and $U = 6\varepsilon$ by directly listing and counting the microstates. Check your work using equation T4.7.

T4B.3 Calculate the multiplicity of an Einstein solid with $N = 1$ and $U = 5\varepsilon$ by directly listing and counting the microstates. Check your work using equation T4.7.

T4B.4 Use equation T4.7 to calculate the multiplicity of an Einstein solid with $N = 4$ and $U = 10\varepsilon$.

T4B.5 Use equation T4.7 to calculate the multiplicity of an Einstein solid with $N = 3$ and $U = 15\varepsilon$.

T4B.6 How many times more likely is it that the combined system of solids described in Table T4.2 will be found in macropartition 3:3 than in macropartition 0:6, if the fundamental assumption is true?

T4B.7 How many times more likely is it that the combined system of solids described in Table T4.2 will *not* be found in macropartition 3:3 than it is to be found in macropartition 0:6, if the fundamental assumption is true?

SYNTHETIC

T4S.1 Explain carefully in your own words the difference between an object's macrostate and its microstate. Why are there generally many microstates in a macrostate?

T4S.2 When two systems are put in thermal contact, how do we describe the macropartition of the combined system? How is this different than the macrostate of the combined system?

T4S.3 Consider an Einstein solid consisting of $N_A = 1$ atom (3 oscillators). Each oscillator can store any integer number of energy units ε. The following table lists the number of microstates g_A available to the solid when it has various values of total thermal energy U_A:

U_A:	0	1ε	2ε	3ε	4ε	5ε	6ε	7ε	8ε	9ε
g_A:	1	3	6	10	15	21	28	36	45	55

By actually listing and counting the various possible microstates, verify the results for the multiplicity g_A for the cases where $U_A = 4\varepsilon$ and $U_A = 7\varepsilon$. (You can check your results using equation T4.7.)

T4S.4 Consider an Einstein solid consisting of $N_B = 2$ atoms. The following table lists the number of microstates g_B available to the solid when it has various values of total internal energy U_B:

U_B:	0	1ε	2ε	3ε	4ε	5ε	6ε	7ε	8ε	9ε
g_B:	1	6	21	56	126	252	462	792	1287	2002

(a) By actually listing the various possible microstates, verify the result for $U_B = 2\varepsilon$. **(b)** Using equation T4.7, verify the value of g_B for $U_B = 6\varepsilon$ and 9ε.

T4S.5 Imagine putting the two solids discussed in problems T4S.3 and T4S.4 into thermal contact. Imagine that the resulting combined system is isolated from everything else and that the combined system contains 6 units of energy (that is, $U_A + U_B = 6\varepsilon$). **(a)** Construct a table showing U_A, U_B, g_A, g_B and g_{AB} for all possible macropartitions of the system (i.e. a table analogous to Table T4.2) and compute the probabilities for each of the seven possible macropartitions according to the fundamental assumption. **(b)** What is the most probable macropartition of this system? In this case, the most probable macropartition is *not* the macropartition where the two systems have equal energies. How does the energy stored *per atom* in each solid compare in this macropartition?

T4S.6 Imagine putting the two solids discussed in problems T4S.3 and T4S.4 into thermal contact. Imagine that the resulting combined system is isolated from everything else and that the combined system contains 9 units of energy (that is, $U_A + U_B = 9\varepsilon$). **(a)** Construct a table showing U_A, U_B, g_A, g_B and g_{AB} for all possible macropartitions of the system (i.e. a table analogous to Table T4.2) and compute the probabilities for each of the ten possible macropartitions according to the fundamental assumption. **(b)** What is the most probable macropartition of this system? In this case, the most probable macropartition is *not* the macropartition where the two systems have equal energies. How does the energy stored *per atom* in each solid compare in this macropartition?

T4S.7 Imagine putting two solids with $N_A = N_B = 2$ in thermal contact, and imagine that the resulting combined system is isolated from everything else and that it contains 9 units of energy (that is, $U_A + U_B = 9\varepsilon$). Construct a table showing U_A, U_B, g_A, g_B and g_{AB} for all possible macropartitions of the system (i.e. a table analogous to Table T4.2) and compute the probabilities for each of the ten possible macropartitions according to the fundamental assumption. In this case, is the most probable macropartition the one where the energy is evenly divided between the identical solids?

RICH-CONTEXT

T4R.1 Consider a system of N microscopic magnets in a uniform magnetic field whose strength is B. Assume that each microscopic magnet can be oriented in *only* two directions: either with the field direction (let us call this direction *down*) or oriented opposite to the field (*up*). Assume that each magnet's energy is $-\mu B$ in the first case and $+\mu B$ in the second case, where μ is some constant (the same for each magnet) that expresses the magnet's strength. The magnets do not move and there is no other way for their energy to change except by changing their orientation.

The macrostate of such a system is completely described by the variables B, μ, N and the system's total energy U. This model describes certain kinds of solids at extremely low temperatures where other modes of energy storage are frozen out. (Like the Einstein solid, it is relatively easy to count microstates for this model, but unlike the Einstein solid, this model only describes fairly esoteric systems in the real world, and behaves in certain peculiar ways that do not generalize well to other systems.)
(a) Imagine that in a system of N such magnets, n magnets are oriented up and the remainder are oriented down. Find a formula for U in terms of n, N, μ and B. (Assume the magnets cannot move or change their energy in any way except by changing their orientation with respect to the field.)
(b) Imagine that you have a system of N such magnets with known strength μ in a magnetic field of known strength B and the system has a known energy U, (and thus a known number n of magnets that are oriented up). How many microstates (different sets of individual magnet orientations) correspond to this macrostate? [*Hints*: Play around with small systems first. For example, imagine that we have three magnets and the energy U is consistent with two of these magnets being oriented up. The possible microstates are then ↑↑↓, ↑↓↑, and ↓↑↑. Once you have developed some sense of how it goes for small systems start thinking about a general formula. The formula will have a $2N$ in the numerator and some factorials in the denominator. Once you have a trial formula, go back and check it in small-system cases where you can directly count the microstates.]

ADVANCED

T4A.1 (a) Show that if you take the time-derivative of equation T4.1, you get

$$0 = v_x(ma_x + k_s x) + v_y(ma_y + k_s y) + v_z(ma_z + k_s z) \quad (T4.10)$$

(b) By changing initial conditions, I could arrange it so v_x, v_y, and v_z had any values I pleased at a given time. Since this equation has to be zero at *all* times, no matter what I arrange v_x, v_y, and v_z to be, the quantities in parentheses have to be independently equal to zero at all times. Show that each of these quantities is the same as Newton's second law for a simple one-dimensional harmonic oscillator moving in a given axis direction.

T4A.2 Equation T4.7 can be derived as follows. First of all, note that the problem of counting microstates of an Einstein solid is essentially the same as the problem of finding the number of distinct patterns that can be generated by pulling marbles and matchsticks randomly from a bag. For example, imagine that we pull the following sequence of items from the bag (reading from left to right):

$$O \mid O\,O\,O \mid\mid O\,O \mid O \mid O$$

If we imagine each marble to be a unit of energy and each matchstick to be a *division* between two oscillators, then this pattern corresponds to the 130211 microstate for an Einstein solid with $N = 2$ atoms (6 oscillators) and $U = 8\varepsilon$.

(a) Argue that a solid with M oscillators and n units of energy will be represented by n marbles and $M-1$ matches.

(b) Argue that if we put $M-1$ matches and n marbles in the bag, there will be a total of $(M+n-1)!$ different ways of pulling objects out of the bag. [*Hint*: When we pull out the first object, there are $M+n-1$ choices. When we choose the second item, there are now only $M+n-2$ choices, since we've already pulled out one object, and so on.]

(c) Not all of these distinct ways of pulling out objects generate distinct patterns. For example, consider taking a given pattern and rearranging the marbles. The rearrangement does not change the basic pattern, but would represent a different sequence of choices as we pull objects out of the bag, and thus would be counted as a distinct choice in part b. Argue that there are $n!$ ways of rearranging the marbles and $(M-1)!$ ways of rearranging the matchsticks without affecting the pattern.

(d) Argue, then, that equation T4.7 correctly states the number of distinct patterns that can be generated, and thus the number of distinct microstates of an Einstein solid with M oscillators and n units of energy .

ANSWERS TO EXERCISES

T4X.1 About 1.1×10^9. Yes.
T4X.2 011, 101, 110, 200, 020, 002.
T4X.3 111, 012, 021, 102, 201, 120, 210, 003, 030, and 300, so $g(1,3\varepsilon) = 10$.
T4X.4 $g(3,8\varepsilon) = 12,780$.
T4X.5 The table is as follows:

Macro-partition	U_A	U_B	g_A	g_B	g_{AB}
0:8	0	8	1	45	45
1:7	1	7	3	36	108
2:6	2	6	6	28	168
3:5	3	5	10	21	210
4:4	4	4	15	15	225
5:3	5	3	21	10	210
6:2	6	2	28	6	168
7:1	7	1	36	3	108
8:0	8	0	45	1	45
Grand total number of microstates			=		1287

T4X.6 The probabilities are as follows

Table 4.2:		Exercise T4X.5	
Macro-partition	Probability	Macro-partition	Probability
0:6	6.0%	0:8	3.5%
1:5	13.6%	1:7	8.4%
2:4	19.5%	2:6	13.1%
3:3	21.6%	3:5	16.3%
4:2	19.5%	4:4	17.5%
5:1	13.6%	5:3	16.3%
6:0	6.0%	6:2	13.1%
		7:1	8.4%
		0:8	3.5%

The 4:4 macropartition is the most probable one in the list for situation in Exercise T4X.5.

T5

THE CONCEPT OF ENTROPY

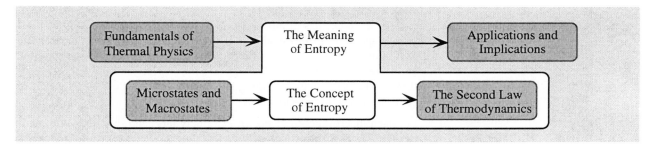

T5.1 OVERVIEW

In the previous chapter, we introduced Einstein's model for a solid and used this model as a context for exploring the most fundamental concepts of statistical mechanics (the *m*-words *macrostate*, *microstate*, *multiplicity*, and *macropartition*, as well as the *fundamental assumption* of statistical mechanics). In this chapter, we continue the process, using a computer to help us prepare macropartition tables for successively larger Einstein solids. This will help us bridge the gap between the very tiny systems (one and two atoms) considered in the last chapter and the larger systems (10^{23} atoms) encountered in daily life.

Here is an overview of this chapter's sections

T5.2 *CONCEPTS OF STATISTICAL MECHANICS* briefly reviews the core ideas of statistical mechanics presented in the last chapter.

T5.3 *THE STATMECH PROGRAM* introduces the computer program that we will use to generate for larger systems the same kinds of macropartition tables that we created in the last chapter by hand.

T5.4 *WHAT HAPPENS AS SYSTEMS BECOME LARGE?* explores how irreversibility naturally arises out of the random shuffling between microstates when the system becomes large.

T5.5 *THE DEFINITION OF ENTROPY* discusses the concept of entropy and how we can use it to quantify irreversible behavior.

T5.6 *EQUILIBRIUM AND AVERAGE ENERGY PER ATOM* investigates systems where the solids in contact are *not* equal in size, using *StatMech* to determine how the energy in such systems is distributed in the most probable macropartition.

*** *MATH SKILLS: LOGARITHMS* reviews the basic properties of logarithms and summarizes their properties. We will be using logarithms extensively from this chapter on.

In a certain sense, this chapter is the center of this unit. It is in this chapter that we begin to engage the puzzle of irreversibility fully, at least in the context of an Einstein solid. We will generalize what we discover to other kinds of systems in chapter T6, and then explore applications and implications of these ideas in chapters T7 through T9.

T5.2   CONCEPTS OF STATISTICAL MECHANICS

In the previous chapter, we developed the basic concepts of the field of physics known as *statistical mechanics.* Using the Einstein solid as the guiding example, we learned that:

1. We describe the *macrostate* of a system by specifying a sufficient number of macroscopically measurable variables such as *P, V, N, T* or *U.* In the case of an Einstein solid, knowing *N* and *U* alone suffices to define the solid's macrostate.

2. We describe the *microstate* of a system by specifying the quantum state of each of its molecules or atoms. In the case of an Einstein solid, the atoms behave quantum mechanically as if they were $3N$ independent harmonic oscillators with the same natural frequency ω and each of which can hold an integer number of units of energy $\varepsilon = \hbar\omega$. Specifying the microstate of the solid boils down to describing how many such units of energy each one of these $3N$ oscillators contains.

3. For each macrostate, there are an extremely large number of possible microstates that are macroscopically indistinguishable versions of that macrostate. The number of microstates that are consistent with a given macrostate is called the *multiplicity g* of the macrostate.

4. We describe the *macropartition* of an isolated system of two objects in thermal contact by describing the individual *macrostates* of the two objects. In the case of two interacting Einstein solids with known values of N_A and N_B, stating the thermal energies U_A and U_B of each solid suffices to describe the macropartition of the combined system.

5. The *fundamental assumption* asserts that each accessible microstate (that is, each microstate consistent with the system's given total energy U) is equally probable: as energy is continually and randomly shifted from atom to atom in the system, each of the system's microstates is as likely to be "visited" as any other microstate.

6. This in turn means that when our system is two objects in thermal contact, the relative probability that a given macropartition is "visited" in the random shuffling of microstates is proportional to the multiplicity of the combined system in that macropartition.

These ideas provide the key to understanding irreversibility and the idea of entropy. The purpose of this chapter is to work through the consequences of the fundamental assumption (still in the context of Einstein solids) to discover how irreversible processes arise out of the random shuffling of microstates.

T5.3 THE *STATMECH* PROGRAM

Program does what we did to prepare Table T4.2

The computer program *StatMech* allows us to rapidly and easily explore the behavior of two Einstein solids (*A* and *B*) in thermal contact. The program allows you to choose any positive integer values for N_A and N_B (the number of atoms in each solid) and *U* (the total energy of the combined system). After a certain interval spent in calculation (which can be fairly long if N_A, N_B and/or *U* are large), the program then displays a table similar to Table T4.2 in the last chapter. This program does *exactly* the same thing that we did by hand in the last chapter, except that it does it *much faster*, and does not get tired or bored even when working out hundreds of macropartitions.

Entering the values of N_A, N_B and total energy U

When the program is started, it displays the window shown in Figure T5.1 (except that the bottom three-quarters of the window will be blank). Near the top of the window, you can see three text boxes labeled **A Atoms**, **B Atoms** and

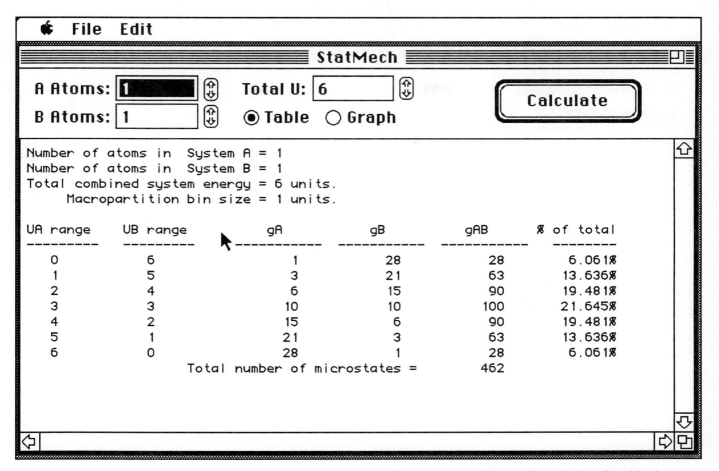

Figure T5.1: Using *StatMech* to display a macropartition table.

Total U, corresponding to the values of N_A, N_B and U respectively. If you press the *tab* key, you can highlight each of these boxes in turn (the first box is highlighted in Figure T5.1), or you can drag over the number in a box using the mouse. When the box is highlighted, you can type in a new value: the computer will accept positive integers between 1 and 1100 inclusive. Alternatively, you can use the mouse to click on either the up-arrow or the down-arrow to the right of each box. Doing this will add or subtract 1 from the number currently in the box. (This works even if the box is not highlighted.)

When you have entered appropriate numbers into these boxes, you can press the **Calculate** button (pressing the *return* key on the keyboard does the same thing). After a brief interval, the program will display a table like the one shown in Figure T5.1. This table lists all the possible macropartitions of the system, and for each macropartition, it displays the multiplicities g_A, g_B and g_{AB} and the percent probability of each macropartition (which, according to the fundamental assumption, is equal to the percent that g_{AB} for each macropartition is of the total number of microstates available to the combined system, which in turn is listed on the last line of the table). At the top of the table, the computer lists (for the record, in case you print the table or copy it to another program) the number of atoms in systems (Einstein solids) *A* and *B*, the total energy of the combined system and the "macropartition bin size" (we will talk more about this shortly).

The table actually displayed in Figure T5.1 shows what happens when the computer is fed the values $N_A = N_B = 1$ and $U = 6\varepsilon$, the same situation for which we prepared Table T4.2. You can see that the computer reproduces Table T4.2 exactly, and even provides the percent probabilities that we painstakingly calculated in exercise T4X.6.

Generating the table

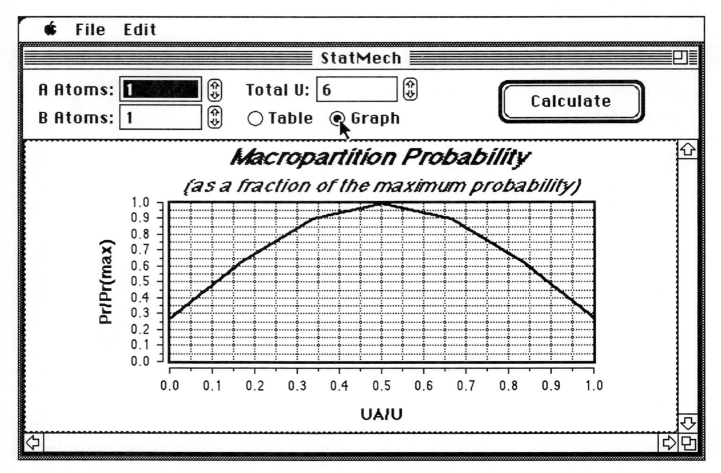

Figure T5.2: Displaying a plot of macropartition probability as a function of the fraction of the total energy that solid *A* has in that macropartition.

If you click on the **Graph** radio button ◉ using the mouse, you will see a graph (see Figure T5.2) of these probabilities plotted versus the fraction U_A/U expressing the energy of solid *A* as a fraction of the combined system's total energy (on the extreme left side of the graph, *A* has no energy, on the extreme right side, *A* has *all* of the combined system's energy).

The vertical scale of this graph shows the probability of a given macropartition as a *fraction* of the probability of the most probable macropartition. This ensures that the peak of the curve will always be at the top of the vertical scale.

Clicking on the **Table** radio button restores the table. You can change the values for N_A, N_B, and U at any time: when you press the **Calculate** button (or hit the *return* key), the computer will calculate a new table (or graph) that overwrites the table (or graph) currently displayed.

Printing the table or graph

You can print either the table or graph as follows. Use the radio buttons to display the table or graph (whichever you want), and then click with the mouse on the table or graph to activate it. Then hold the ⌘ key and press the P key (or select "Print" from the File menu). Hit the return key in response to the printer dialog box that comes up, and your printout should soon come rolling out.

Copying table or graph to another program

To copy all or part of the table to another program, you should use the radio buttons to display the table, hold ⌘ key and press the A key to highlight the entire table (or choose "Select All" from the Edit Menu or drag with the mouse to highlight a subset of the table), and then hold the ⌘ key and press the C key (or select "Copy" from the Edit menu) to copy the table to the computer's internal clipboard. Then transfer to your receiving program and paste the text using whatever commands are appropriate for that program. The process for copying the graph is the same except that you do NOT need to press ⌘A (or choose "Select All"): the entire graph is *automatically* selected.

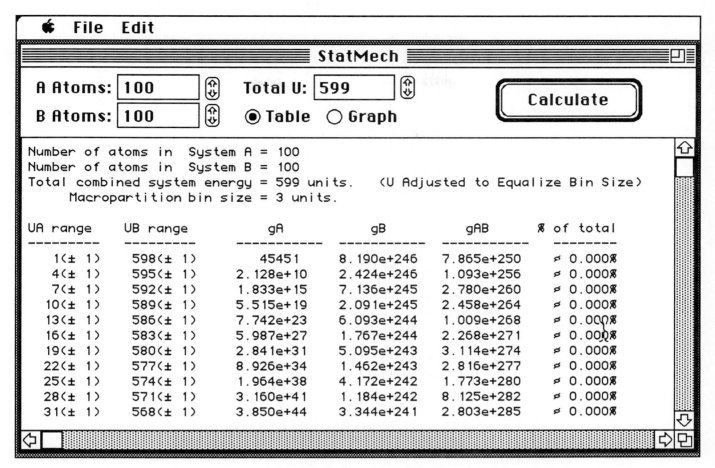

Figure T5.3: A macropartition table showing how the computer groups macropartitions into bins when *U* becomes large.

When *U* becomes large, the program groups macropartitions into "bins"

If your chosen value of *U* exceeds 200 units of energy, rather than print over two hundred lines-worth of distinct macropartitions, the computer begins to cluster macropartitions into groups or *bins*. For example, Figure T5.3 shows what happens when we attempt to scale the system shown in Figures T5.1 and T5.2 up by a factor of 100 (changing N_A and N_B from 1 to 100, and changing *U* from 6 units of energy to 600 units). When you press the **Calculate** button the computer beeps, and then readjusts the energy slightly (in this case from 600 to 599) so that the number of macropartitions is exactly divisible (in this case) by three: (there are then exactly 600 macropartitions corresponding to $U_A = 0$ to 599 after the adjustment). It then lists the macropartitions in clusters or *bins* of three, for a total of 200 bins of three. Note that the computer informs you at the top of the table that "U [has been] Adjusted to Equalize Bin Size" and also lists the number of macropartitions it has put in each bin.

Under these circumstances, each line of the computer-generated table corresponds to a bin instead of a single macropartition. For example, the first line in the table shown in Figure T5.3 describes a bin of three macropartitions whose U_A values in the range 1±1 unit (that is, 0, 1, or 2 units). The multiplicities in each line are *sums* for the macropartitions within that range.

(Note that in this case, the probabilities for the macropartition bins visible in the window are "≈ 0.000%". The probabilities are not strictly zero, of course, but are less than 0.001% and so cannot be displayed within the format the computer uses in this column. Also note that the table is too long to be entirely displayed within the window: to display the lower parts of the table, you can use the mouse to press on the down-arrow in the lower right corner of the window. You can also use the down-arrow on your keyboard, but this is much slower. A number displayed as 7.865e+250 means 7.865×10^{250}.)

Grouping macropartitions into bins is partly to keep the length of the tables manageable, but also reflects something important about physical practicalities in a case like this. Since a macropartition is defined in terms of the macrostates of the two interacting solids, it is only as well-defined as the macrostates. Imagine that macroscopic measurements of the thermal energy of an Einstein solid are only good to about ±1% of the fixed combined energy of the system. If the energy difference between two hypothetical adjacent macropartitions is less than this threshold, they cannot be distinguished by macroscopic measurements and thus do not really represent two *experimentally* distinct macropartitions. Any practical measurement of the two solids' energies can only therefore specify a *range* of theoretical macropartitions within which the true macropartition will be found. By grouping theoretical macropartitions into bins as U gets large, the program is just doing what we would have to do in a realistic situation anyway.

The preceding pages describe version 1.0a of *StatMech*. That version is limited to $N_A \leq 1100$, $N_B \leq 1100$, and $U \leq 9995$ units (numbers beyond these values produce multiplicities so large that they overflow the computer's calculating unit). Later versions may have additional features and capabilities that your instructor will describe if you are using such a version.

Exercise T5X.1: If you have *StatMech* on a computer that you can use while reading the rest of this chapter, you should go to that computer now. Test the features described in this section, and check the table provided in the answer for Exercise T4X.5 in the last chapter. Then try running cases where you gradually scale up the size of the system that we imagined when we constructed Table T4.2. There we had $N_A = N_B = 1$ and $U = 6\varepsilon$. Use *StatMech* to investigate what happens when we multiply these numbers by 10, 30, 100, and 1000. What happens to the multiplicities in the tables? What (qualitatively) happens to the graphs of macropartition probabilities?

[Note: You do not *have* to have access to a computer to finish reading this chapter. Everything you need to know will be described. The ideas will, however, become more vivid and memorable if you do use the program.]

T5.4 WHAT HAPPENS AS SYSTEMS BECOME LARGE?

Program enables us to study larger systems

In the last chapter, we were forced to consider interacting Einstein solids consisting of at most two or three atoms and sharing only a handful of units of energy, because as these numbers get larger, it rapidly becomes impossible to do the calculations by hand. Yet we know that even the smallest dust mote of a solid contains thousands of trillions of atoms. It turns out that the irreversible nature of macroscopic processes depends crucially on the fact that an incredibly huge number of atoms are involved in a macroscopic process.

The *StatMech* program allows us to explore the behavior of larger solids than we can calculate by hand, though the number of atoms that the program can handle is still far smaller than any real solids would contain. Still, by running the program with an increasing number of atoms involved in each solid, we can spot trends that allow us to extrapolate to even larger numbers.

As system becomes large, multiplicities become *huge*

If you did the runs suggested in Exercise T5X.1, you should have noted several things. First of all, it should have become apparent that as N_A, N_B, and U get large, the number of accessible microstates goes *through the roof*. In the case where $N_A = N_B = 30$, $U = 180\varepsilon$ (microscopically small solids only about three atoms on a side!), the total number of microstates available to the combined system is on the order of magnitude of 5×10^{106}, which is bigger than Avogadro's number times the total number of protons, neutrons, and electrons in the universe. This is already an unimaginably large number, and things only get more incredible as N_A, N_B, and U get larger: if $N_A = N_B = 1000$ and $U = 6000\varepsilon$ (these

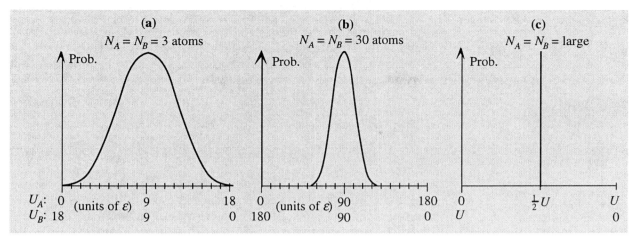

Figure T5.4: How the graph of macropartition probability graph changes as the size of a system get larger.

are still tiny solids only 10 atoms on a side), the total number of microstates accessible to the solid is on the order of 5×10^{3599}, which (if I did not use scientific notation) would take more than a page of this text just to write out all the zeros! It turns out that these extremely large numbers play an important role in making processes irreversible, as we will soon see.

Secondly, you should have noticed that if we keep $N_A = N_B$, the most probable macropartition in each run is the one where the energy is split evenly between the two identical solids. The *shape* of the probability graph, however, changes as N_A, N_B, and U get larger. When N_A, N_B, and U are small (Figure T5.4a), the graph of probability versus macropartition is a fairly broad bell-shaped curve, meaning that even the lowest-probability macropartitions have a significant probability of being seen if you watch the combined system long enough. As N_A, N_B, and U get larger, though, the curve gets narrower (Figure T5.4b) Perhaps you can imagine that as these numbers get very large, the probability graph narrows to a spike of almost infinitesimal width.

As the system becomes large, the probability graph becomes more sharply peaked

My claim is that these facts have two very important implications:

Implications

1. If the combined system is *not* in the most probable macropartition to begin with, it will rapidly and inevitably move toward that macropartition, and

2. It will subsequently *stay* at that macropartition (or very near to it), in spite of the random shuffling of energy back and forth between the two solids.

I claim in fact that these statements follow directly from the huge number of microstates involved and the narrowness of the probability graph. How?

Let's consider the second implication first. Imagine that our two Einstein solids have nearly the same energy before they are brought into contact. After they are brought into contact, they randomly shuffle energy around internally and across the boundary between them, sampling various possible microstates. The fundamental assumption implies that the probability that the system will end up in a given microstate after one of these shuffles is the same for all microstates, and since some microstates are in different macropartitions, the combined system's macropartition will **fluctuate** randomly in time.

Discussion of the second implication

With a small system (say $N_A = N_B = 3$ atoms, $U = 18\varepsilon$), these fluctuations can be pretty significant. Even if the system starts out near the center macropartition, there is about a 0.03% probability that if we peek at the system at some later time we will find it to have wandered into one of the extreme macropartitions (where all the energy is in one solid and none in the other). Fluctuations of $\pm 4\varepsilon$ (which corresponds to almost $\pm 50\%$ of the most probable value 9ε of the energy in each solid) will be seen quite commonly.

On the other hand, even in only a modestly larger system (with $N_A = N_B = 30$ atoms and $U = 180\varepsilon$, for example), the probability graph narrows significantly, and the probability of observing large fluctuations gets much smaller. For example, the probability of observing this system to be in its most extreme observable macropartition (where $U_A \approx 0$ and $U_B \approx 180\varepsilon$) turns out to be:

$$\frac{g_{AB} \text{ in extreme macropartition}}{\text{grand total } g_{AB}} \approx \frac{7 \times 10^{72}}{5 \times 10^{106}} \approx 1.4 \times 10^{-34} \qquad (\text{T5.1})$$

This is such an extremely small number that is really zero in any practical sense (see Exercise T5X.3). Fluctuations in this system's macropartition will essentially be confined to a range of about ±20% of the most probable energy of each solid. If we increase the size of the system to $N_A = N_B = 1000$, $U = 6000\varepsilon$, we find that more than 99% of the available microstates are in macropartitions where the objects' energies are within ±10% of their most probable values.

So, if two solids are initially at or near the central macropartition, then, they will not stray very far from that macropartition. The fundamental reason for this is that *the vast majority of microstates are in macropartitions close to the most probable one*. If each microstate is really equally probable, the system will seem to strongly prefer being in these central macropartitions.

Exercise T5X.2: If you can, run the *StatMech* program and verify that the values of g_{AB} quoted in equation T5.1 are roughly correct. Estimate the probability that the system will be found in the 1:179 macropartition.

Exercise T5X.3: The age of the universe is probably about 1.5×10^{10} y, and there are about 3×10^7 s in a year. If you have been peeking at a system of two Einstein solids with $N_A = N_B = 30$ atoms and $U = 180\varepsilon$ a billion times a second from the day the universe began, roughly what would be the probability that you would have seen the system in its most extreme observable macropartition (0:180) by now? (The point is that even though it is *theoretically* possible that this macropartition might occur, it is *impossible* in any practical sense.)

Discussion of the first implication: system moves inexorably toward the center

What if you begin with the solids in a fairly extreme macropartition? Once the solids are brought into thermal contact, they will begin to randomly shuffle energy back and forth, sampling new microstates. But the vast majority of microstates near to the original macropartition are to be found in the direction toward the central macropartition, so the macropartition of the system will almost certainly move in that direction.

For example, consider the numbers shown in Figure T5.3. Let us imagine that we start in the macropartition where $U_A = 10\varepsilon$ (±1ε) and $U_B = 589\varepsilon$ (±1ε): this would correspond to solid A having a relatively low temperature and solid B having a relatively high temperature. The multiplicity of this macropartition is about 2.5×10^{264}. The multiplicity of the adjacent macropartition corresponding to a more even split of energy, where $U_A = 13\varepsilon(\pm1\varepsilon)$ and $U_B = 586\varepsilon(\pm1\varepsilon)$, is about 1.0×10^{268}, about 4,000 times larger. The multiplicity of the adjacent macropartition corresponding to a more *extreme* distribution of energy is about 2.8×10^{260}, nearly 10,000 times smaller than the multiplicity of the original macropartition. If in a given short period of time the system is likely to evolve from its initial microstate to microstates within the original or adjacent macropartitions, we see from these numbers that it is about 4,000 times *more* likely that the system will evolve toward a more even distribution of energy than remain with the same distribution, and 10,000 times *less* likely to evolve by chance to a more extreme distribution. If the system changes macropartition, there is only about one chance in 40 million that the change will be in the direction *away* from a more even distribution of energy between the objects.

As the number of atoms involved becomes even larger, this probability becomes a virtual *certainty*. In a system where the solids have $N_A = N_B = 1000$ atoms and we start out in the macropartition where $U_A = 127\varepsilon(\pm25\varepsilon)$ and $U_B = 5839\varepsilon(\pm25\varepsilon)$, the system is 4.7×10^{54} times *more* likely to evolve toward a more even distribution of energy than stay in the same macropartition and about 8.6×10^{61} times *less* likely to evolve toward a macropartition with a less even distribution of energy than stay in the original one. When the numbers become this large, the idea that it is possible in *principle* for the system to stay in the same macropartition (or even move to a more extreme one) has no *practical* meaning: we will never, never, *never*, even if we watch for Avogadro's number of lifetimes of the universe, observe this system to do anything but inexorably march toward macropartitions where the energy is more evenly distributed.

Exercise T5X.4 Here is an excerpt from the table produced by *StatMech* for the case discussed above. Check the numbers quoted in the paragraph above. If you can run *StatMech*, check the excerpt also. Note that the solids involved here are *still* infinitesimal specks (10 atoms on a side!) compared to normal objects.

U_A range	U_B range	g_{AB}	
25(±25)	5941(±25)	1.182e+2580	for:
76(±25)	5890(±25)	2.279e+2653	$N_A = 1000$
127(±25)	5839(±25)	1.950e+2715	$N_B = 1000$
178(±25)	5788(±25)	9.179e+2769	$U = 5966\varepsilon$

Let's take a step back and consider what this means at the macroscopic level. I take a solid with a lot of thermal energy and place it in contact with an otherwise identical solid with little thermal energy. What we see is that once contact is established, the random shuffling of energy between the objects leads with *virtual certainty* to a net energy flow from the high-energy solid to the low-energy solid until they reach an even energy distribution. Once the solids reach that central macropartition, they will stay there, again with virtual certainty, except for tiny (perhaps even immeasurable) random fluctuations.

The link to paradigmatic thermal process

Now, two solids with high and low thermal energies would normally be considered to have high and low temperatures respectively. So what we are seeing here is that when a high-temperature solid is brought into contact with a low-temperature (but otherwise identical) solid, energy flows (on the average) from the hot solid to the cold solid until they reach an equilibrium state where their thermal energies are equal. Subsequently, they remain in that state. Does this sound familiar?

This, then, (at least in the special context of identical Einstein solids) is the answer to the fundamental question of this unit. Why are some macroscopic processes irreversible? Specifically, why does heat always flow from hot to cold and never the other way? *Because there are more microstates in that direction.* Not because energy cannot in *principle* flow the other way. Not because the time-reversible laws of microscopic physics for some reason don't apply here. Complex objects simply evolve by random process to the macroscopic states that contain the most microstates. It is really (surprisingly, wonderfully) simple!

Look back over the argument that I have presented in this section. The details of Einstein solid and exactly how many microstates they have in this macropartition as opposed to that macropartition are not really very important. There are really only two things that have to be true for this explanation of irreversibility to work:

The argument presented here is fairly general

1. The number of microstates available to the interacting solids must be *huge*,
2. The graph of probability as a function of macropartition must have a very narrow peak.

Exercise T5X.5: Why is it necessary that the number of available microstates be huge? What effect does this have on the comparative probabilities of various macropartitions? Why must the probability have a very narrow peak? What does this mean for the size of random fluctuations? Think about these issues and then briefly answer these questions as best you can in your own words.

T5.5 THE DEFINITION OF ENTROPY

To know which macropartitions of a combined system are the most probable, we need to know how many microstates the system has in that macropartition. However, as we've seen, the number of microstates available to even a tiny system can be so large as to be awkward to deal with. The number of microstates available to Einstein solids involving Avogadro's number of atoms can be of the order of magnitude of $10^{10^{23}}$ (that is, one followed by Avogadro's number of zeros). Such numbers are truly unmanageable. For realistically sized objects, we need a less awkward way to talk about multiplicity.

Toward this end, we define the **entropy** S of any object in a given macro-state to be:

The definition of entropy

$$S \text{ (of the macrostate)} \equiv k_B \ln g \text{ (of the macrostate)} \tag{T5.2}$$

where k_B is Boltzmann's constant and g is the multiplicity of that macrostate (that is, the number of microstates whose characteristics are consistent with the specification of that macrostate). In the specific case of an Einstein solid, whose macrostate is specified by U and N, its entropy is a function of U and N:

Entropy of Einstein solid

$$S(U,N) = k_B \ln g(U,N) = k_B \ln \left[\frac{(3N + U/\varepsilon - 1)!}{(3N - 1)!(U/\varepsilon)!} \right] \tag{T5.3}$$

Some nice features of this definition

Why this definition? *Entropy* is really just another way of talking about the *multiplicity*: when the multiplicity is large, the entropy is large. Defining the entropy in terms of the *natural logarithm* of the multiplicity, however, serves two very useful purposes. First of all, it makes the awkwardly large multiplicity values easier to manage. For example, the logarithm of a multiplicity like $g = 10^{10^{23}}$ is about 2.3×10^{23}, which while still a large number, is much more manageable than the multiplicity itself.

Secondly, consider a specific macropartition of a system of two Einstein solids in thermal contact. The multiplicity of the combined system in this macropartition is $g_{AB} = g_A \cdot g_B$, where g_A and g_B are the multiplicities of the individual solids in their specified macrostates.

Exercise T5X.6: One of the basic properties of the natural logarithm function is that $\ln(xy) = \ln x + \ln y$. Show that this means that the total entropy of the *combined* system in a given macropartition is

$$S_{AB} = S_A + S_B \tag{T5.4}$$

In other words, the entropy of the *combined* system known to be in a given macropartition is the sum of the entropies of the individual subsystems.

The fact that entropies of two systems in contact *add* (instead of multiply, as multiplicities do) is a desirable outcome of our definition of entropy.

The constant k_B just rescales the logarithm in a way that we will see is natural and helpful when we explore the definition of temperature in the next chapter. For right now, you can simply note that since $k_B = 1.38 \times 10^{-23}$ J/K a multiplicity on the order of magnitude of $10^{10^{23}}$, and thus whose logarithm is about 2.3×10^{23}, will have a corresponding entropy of about 4 J/K, a number of convenient size.

In the previous section, we argued that a system of two otherwise identical Einstein solids with different internal energies will evolve from macropartitions with lower multiplicity (and thus lower entropy) to macropartitions with higher multiplicity (higher entropy). Moreover, once the system reaches the macropartition having the highest multiplicity (highest entropy), it will stay there. Therefore, as the system evolves from macropartitions with low multiplicity to those with larger multiplicity, *its entropy increases*. The entropy of a (sufficiently large) system will never be observed to decrease, because the probability that the system would evolve to a macropartition where its entropy is measurably smaller is extraordinarily improbable (see problem T5R.2 for a numerical example). This is our first glimpse of the **second law of thermodynamics**: *The entropy of an isolated system never decreases.*

The second law of thermodynamics

Problem: In the situation where we have two Einstein solids in thermal contact and $N_A = N_B = 100$ and $U = 599\varepsilon$. (a) Find the entropies of both individual solids and the total system in the macropartition where $U_A = 13(\pm 1\varepsilon)$. (b) Compare the entropy of the total system in this macropartition with that of the total system in the two central macropartitions (where $g_{AB} = 6.8 \times 10^{357}$). It is acceptable to express these entropies as a multiple of k_B.

EXAMPLE T5.1

Solution According to Figure T5.3, $g_A = 7.74 \times 10^{23}$, $g_B = 6.09 \times 10^{244}$, and $g_{AB} = 1.01 \times 10^{268}$. The first of the entropies is easy to calculate directly: $S_A = k_B \ln g_A = 55.0 k_B$. However, my calculator cannot handle exponents greater than 99, so I cannot calculate $\ln g_B$ directly. However, $\ln(x^a) = a \ln x$, so this means that $\ln(10^{244}) = 244 \cdot \ln 10 = 244(2.30258)$. Moreover, $\ln(xy) = \ln x + \ln y$, so

$$\ln g_B = \ln(6.09) + \ln(10^{244}) = 1.807 + 244(2.30258) = 563.6 \qquad (T5.5)$$

so $S_B = 563.6 k_B$. Similarly $S_{AB} = k_B[\ln(1.01) + 268(2.30258)] = 617.1 k_B$.[*] The entropy of the central macropartitions is $S_{AB} = k_B[\ln(6.8) + 357(2.30258)] = 823.9 k_B$, which is substantially larger.

Problem: Imagine that one macropartition of a combined system of two Einstein solids has an entropy of $432.5 k_B$, while another (where the energy is more evenly divided) has an entropy of $546.3 k_B$. How many times more likely are you to find the system in the second macropartition compared to the first?

EXAMPLE T5.2

Solution The probability of a given macropartition being observed is proportional to its multiplicity g. According to the definition of entropy $S \equiv k_B \ln g$, so (since $e^{\ln x} = x$) $g = e^{S/k_B}$. Therefore, the ratio of the probabilities in this case is

$$\frac{\text{Prob}(mp_2)}{\text{Prob}(mp_1)} = \frac{g_2}{g_1} = \frac{e^{S_2/k_B}}{e^{S_1/k_B}} = \frac{e^{546.3}}{e^{432.5}} = e^{546.3-432.5} = 2.6 \times 10^{49} \qquad (T5.6)$$

where mp_1 and mp_2 refer to the macropartitions. Therefore, we are 2.6×10^{49} times more likely to find it in the second rather than the first macropartition.

---

[*]This is slightly smaller than the sum $S_A + S_B = 618.6 k_B$ because g_{AB} in the table shown in Figure T5.3 is not actually the product of g_A and g_B as listed (as you can easily check). Rather, it is the *sum* of the products $g_A g_B$ for the three actual macropartitions embraced by that line in the table, just as g_A is the *sum* of the g_A values and g_B is the *sum* of the g_B values for the same three macropartitions. We should not expect the *sum of the products* $g_A g_B$ represented by g_{AB} to be exactly the same as the *product of the sums* represented by g_A times g_B. The discrepancy in the entropies decreases as the numbers get larger and as we get closer to the central macropartition.

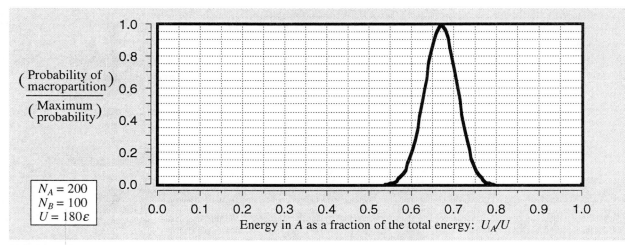

$N_A = 200$
$N_B = 100$
$U = 180\varepsilon$

Energy in A as a fraction of the total energy: U_A/U

Figure T5.5: Graph of macro-partition probability as a function of energy distribution for a pair of unequal solids.

The definition of the equilibrium macropartition

The average energy per atom is roughly the same in equilibrium

T5.6 THE AVERAGE ENERGY PER ATOM

As we have seen, a macroscopic high-energy Einstein solid placed in thermal contact with a low-energy (but otherwise identical) Einstein solid will almost inevitably give up energy to the latter until the system reaches an **equilibrium macropartition** (defined to be the most probable macropartition). Once the system has reached this most probable macropartition, the individual solids' thermal energies will no longer change (except for tiny random fluctuations).

In this symmetric case, the equilibrium macropartition is the one where the two solids' total energies are equal. But, what if the solids are *not* the same size? Imagine, for example, that solid A is twice as big as solid B. Figure T5.5 shows a *StatMech* graph for $N_A = 200$, $N_B = 100$, and $U = 180\varepsilon$. According to the graph, the most probable macropartition here is *not* where the solids have the same energy, but rather where $U_A/U = 0.67$, meaning that the larger solid has twice as much energy as the smaller solid B. If we look at the table, we find that the most probable macropartition is indeed the one where $U_A = 120\varepsilon$ and $U_B = 60\varepsilon$. In this case, we thus see that *the system's equilibrium macropartition is the one where each solid gets the same average energy per atom* (0.60ε here).

Exercise T5X.7: Is this *generally* true? If you have access to *StatMech*, try some other unequal systems. It is generally easiest to find the *precise* most probable macropartition if you keep $U \leq 200\varepsilon$, but you should try at least *some* larger numbers for U as well as different values for the ratio N_A/N_B. Is the statement given above always true? Is it approximately true? What can you find out?

The link between energy per atom and temperature

The evidence that you gathered in the previous exercise should indeed support the hypothesis that the equilibrium (most probable) macropartition is the one where the average energy per atom in each solid is the same. Remember that the most basic definition of temperature is "a quantity that is equal when two objects are in thermal equilibrium". In this case, we are seeing that in the case of Einstein solids (at least) it is *the average thermal energy per atom* that becomes equal for two solids when they come into equilibrium. This suggests that the average thermal energy per atom must be linked to an object's temperature, just as we found in the case of the ideal gas in chapter T2. The difference is that in chapter T2 we had to use the standard definition of temperature in terms of the constant volume gas thermometer to arrive at this result. In this case, we have not had to appeal to *any* definition of a temperature scale: the link between temperature and average energy per atom pops out of our observed results and the most basic physical meaning of temperature. We'll explore this issue in more generality in the next chapter.

I. THE CONCEPTS OF STATISTICAL MECHANICS
   A. Concepts discussed in the last chapter
      1. A *macrostate* is specified by macroscopic variables like U and N
      2. A *microstate* is specified by stating each molecule's quantum state
      3. The *multiplicity g* is the number of microstates in a macrostate
      4. A *macropartition* is specified by describing the macrostates of a set of objects in thermal contact
      5. The *fundamental assumption* states that *all accessible microstates are equally likely*. This means that the probability of seeing a system in a given macropartition is proportional to number of microstates available to the system in that macropartition.
   B. This is sufficient to explain irreversibility, as we will see

II. THE STATMECH PROGRAM
   A. The program's basic function
      1. You provide N_A, N_B, and total U for two Einstein solids in contact
      2. It provides table of macropartitions like Table T4.2, as well as a list (or graph) of macropartition probabilities computed according to the fundamental assumption
   B. Other features
      1. You can print either table or graph
      2. You can copy and paste either table or graph into another program
      3. It combines macropartitions in bins when U becomes large

III. WHEN THE SYSTEM BECOMES LARGE
   A. As we increase the values of N_A, N_B, and U, the following happens
      1. The multiplicities of macropartitions become *extremely* large
      2. The probability as a function of U_A becomes very sharply peaked
   B. Implications
      1. If the system is *not* in most probable macropartition, there are *far* more microstates in that direction than away, so the system moves inexorably toward that macropartition.
      2. When system is at (or near) the most probable macropartition, the *vast* majority of system's microstates are nearby, so as energy is shuffled around, the macropartition *may* fluctuate, but the sharp peak means that any fluctuations will be very tiny.
   C. So why does energy flow from hot to cold and not the other way?
      1. The system evolves toward macropartitions with most microstates
      2. Macropartitions where energy is evenly distributed just have more!
   D. The argument applies not just to Einstein solids, but whenever
      1. the number of microstates available to a system is huge, and
      2. the graph of probability vs. macropartition is sharply peaked

IV. DEFINITION OF ENTROPY
   A. The *entropy S* (of a macrostate) $\equiv k_B \ln g$ (of the macrostate)   (T5.2)
   B. Desirable features of this definition
      1. Typical multiplicities g are awkwardly large; $\ln g$ is more reasonable
      2. The total entropy S_{AB} of two objects $= S_A + S_B$
   C. *Second law of thermodynamics*: S of an isolated system never decreases

V. EQUILIBRIUM AND AVERAGE ENERGY PER ATOM
   A. The *equilibrium macropartition* is the most probable macropartition
   B. When two interacting objects do *not* have the same size:
      1. The equilibrium macropartition is where objects have the same average energy per atom (quite generally).
      2. But since temperature is defined to be a quantity that is equal in equilibrium, maybe T is linked to average energy per atom...?

MATH SKILLS: Logarithms

Starting in this chapter, we will be using logarithms extensively. The purpose of this section is to review the mathematical properties of logarithms.

The **natural logarithm** function $\ln x$ is defined to be the inverse of the exponential function e^x. Similarly the **base-ten logarithm**, usually written $\log x$ (but sometimes $\log_{10} x$) is defined to be the inverse of 10^x. Therefore

$$e^{\ln x} = x, \quad 10^{\log x} = x, \quad \ln(e^x) = x, \quad \log(10^x) = x \tag{T5.7}$$

Note that $1/e^a = e^{-a}$, $e^a e^b = e^{a+b}$, $e^a/e^b = e^a e^{-b} = e^{a-b}$, and $(e^a)^y = e^{ay}$. If we substitute $a = \ln x$ and $b = \ln y$ into these expressions and take the logarithm, we find that

$$\ln(1/x) = -\ln x \tag{T5.8a}$$

$$\ln(xy) = \ln x + \ln y, \quad \ln(x/y) = \ln x - \ln y \tag{T5.8b}$$

$$\ln(x^y) = y \ln x \tag{T5.8c}$$

Note in particular that $\ln x + \ln y \neq \ln(x+y)$! Similarly, $1/10^a = 10^{-a}$, $10^a 10^b = 10^{a+b}$ and so on, so equations T5.8 *also* apply to base-ten logarithms.

We can convert from base-10 logs to natural logs and vice versa as follows. If we set $x = 10$, take the exponent of equation T5.8c, and define $\log z = y$ ($z = 10^y$), we get

$$10^y = e^{(\ln 10)y} \quad \Rightarrow \quad \ln z = (\ln 10)\log z = 2.3026 \log z \tag{T5.9a}$$

$$\Rightarrow \quad \log z = \ln z / \ln 10 = 0.43429 \ln z \tag{T5.9b}$$

Finally:

$$\frac{d}{dx} \ln x = \frac{1}{x} \tag{T5.10}$$

GLOSSARY

fluctuation: a random change in a combined system's macropartition due to random energy transfers while the system is near equilibrium. (Such fluctuations are possible because macropartitions *very* near the most probable one have nearly the same number of microstates, and thus are about equally probable.)

entropy S (of a macrostate): defined to be $S = k_B \ln g$, where g is the multiplicity of the macrostate. The entropy of a two-object system in a given macropartition is the sum of the entropies corresponding to each object's macrostate in that macropartition: $S_{AB} = S_A + S_B$.

The second law of thermodynamics: the entropy of an isolated system never decreases. In the language of this chapter, this is because the probability of a large system evolving by chance from one macropartition to another with a significantly smaller multiplicity (and thus entropy) is so small as to be zero for all practical purposes.

equilibrium macropartition: another name for *the most probable macropartition*. Two systems in thermal contact will inexorably evolve toward this macropartition and then remain in that macropartition (or its very near neighbors) indefinitely.

TWO-MINUTE PROBLEMS

T5T.1 A system consisting of two interacting parts can never evolve to a configuration where the total system entropy is smaller than it was originally (T or F).

T5T.2 A hot object is placed in contact with a cold object. Heat is observed to flow spontaneously from the hot to the cold object, but not the other direction. According to the argument in this chapter, this is because
A. This increases the entropies of both objects
B. This decreases the entropy of the combined system
C. The system will tend to evolve toward macropartitions that have more microstates
D. All of the above   F.   A and C

T5T.3 Imagine that you discover a strange substance whose multiplicity is always 1, no matter how much energy you put into it. If you put a very large amount of energy into an object (object A) made of this substance, and place it into thermal contact with an Einstein solid (object B) having the same number of atoms but much less energy, what will happen?

A.   Heat flows from A to B until A has virtually no energy
B.   Heat flows from B to A until B has virtually no energy
C.   Heat flows from A to B until they have the same energy
D.   No energy will flow between A and B at all
E.   something else (describe)

T5T.4 Consider a system comprised of two or more objects in contact. The system's total entropy in a given macropartition is always equal to the sum of the entropies of its parts (T or F).

T5T.5 $10^{62}/10^{60} = 100$. What is $\ln(10^{62})/\ln(10^{60})$? (Select the closest response.)
A.   100   B.   10   C.   4.6   D.   1.0

T5T.6 The entropy of a certain macropartition is $102 k_B$. The entropy of another macropartition is $204 k_B$. How much more likely is the system to be in the second macropartition than the first? (Select the closest response)
A.   2 times   D. $e^{102} = 2 \times 10^{44}$ times
B. $e^2 = 7.4$ times   E. $e^{102 k_B}$ times
C. $10^2 = 100$ times

HOMEWORK PROBLEMS

BASIC SKILLS

T5B.1 Many calculators cannot handle exponents greater than 99. How can we calculate a ratio like 5.0×10^{256} divided by 2.0×10^{232}? Remember that

$$\frac{5.0 \times 10^{256}}{2.0 \times 10^{232}} = \frac{5.0}{2.0}\left(\frac{10^{256}}{10^{232}}\right) = \frac{5.0}{2.0} \times 10^{256-232}$$

$$= 2.5 \times 10^{24} \qquad (\text{T5.11})$$

Using the same technique, show that 3.4×10^{375} divided by 8.3×10^{343} is 4.1×10^{31}. Calculate 5.6×10^{835} divided by 3.9×10^{822}.

T5B.2 Consider the system consisting of a pair of Einstein solids in thermal contact. A certain macropartition has a multiplicity of 3.7×10^{1024}, while the total number of microstates available to the system in all macropartitions is 5.9×10^{1042}. If we look at the system at a given instant of time, what is the probability that we will find it to be in our certain macropartition? (See problem T5B.1.)

T5B.3 Consider the system consisting of a pair of Einstein solids in thermal contact. A certain macropartition has a multiplicity of 1.2×10^{346}, while the total number of microstates available to the system in all macropartitions is 5.9×10^{362}. If we look at the system at a given instant of time, what is the probability that we will find it to be in our certain macropartition? (See problem T5B.1.)

T5B.4 Consider the system consisting of a pair of Einstein solids in thermal contact. Imagine that it is initially in a macropartition that has a multiplicity of 8.8×10^{123}. The adjacent macropartition closer to the equilibrium macropartition has a multiplicity of 4.2×10^{134}. If we look at the system a short time later, how many times more likely is it to have moved to the second macropartition than to have stayed with the first? (See problem T5B.1.)

T5B.5 Consider the system consisting of a pair of Einstein solids in thermal contact. Imagine that it is initially in a macropartition that has a multiplicity of 7.6×10^{3235}. The adjacent macropartition closer to the equilibrium macropartition has a multiplicity of 4.1×10^{3278}. If we look at the system a short time later, how many times more likely is it to have moved to the second macropartition than to have stayed with the first? (See problem T5B.1.)

T5B.6 Use *StatMech* to generate a macropartition table for a pair of Einstein solids with $N_A = 20$, $N_B = 40$, and total energy $U = 120\varepsilon$. Hand in a printout of the table, and circle the equilibrium macropartition. What is the average energy per atom for each solid in this macropartition?

T5B.7 Use *StatMech* to generate a macropartition table for a pair of Einstein solids with $N_A = 150$, $N_B = 50$, and total energy $U = 100\varepsilon$. Hand in a printout of the table, and circle the equilibrium macropartition. What is the average energy per atom for each solid in this macropartition?

T5B.8 An Einstein solid in a certain macrostate has a multiplicity of 3.8×10^{280}. What is its entropy (expressed as a multiple of k_B)?

T5B.9 A pair of Einstein solids in a certain macropartition have multiplicities of 4.2×10^{320} and 8.6×10^{132} respectively. What are the entropies of each solid? What is

the total entropy of the system in this macropartition? (Express your entropies as multiples of k_B.)

SYNTHETIC

T5S.1 Is it *really* true that the entropy of an isolated system consisting of two Einstein solids *never* decreases? (Consider a pair of very small solids.) Why is this statement more accurate for large systems than small systems? Explain in your own words.

T5S.2 Run the *StatMech* program for two Einstein solids in contact with $N_A = N_B = 100$ and $U = 200\varepsilon$. **(a)** How many times more likely is the system to be found in the center macropartition than in the extreme macropartition where $U_A = 0$ and $U_B = 200\varepsilon$? **(b)** What is the range of values that U_A is likely to have more than 95% of the time? **(c)** If U_A were initially to have the extreme value 0, how many times more likely is it to move to the next macropartition nearer the center than remain in the extreme one?

T5S.3 (Do Problem T5S.2 first.) Repeat Problem T5S.2 for the case of two Einstein solids in contact, but scale everything up by a factor of 10, so that $N_A = N_B = 1000$ and $U = 2000\varepsilon$. (Treat each line of the table as if it described a single macropartition.) Compare your answers with those that you found in Problem T5S.2.

T5S.4 (Do Problem T5S.2 first.) Repeat Problem T5S.2 except increase the number of atoms by a factor of 10 (so that $N_A = N_B = 1000$) but do not change U ($U = 200\varepsilon$ still). Compare your answers with those that you found in Problem T5S.2. Comment on how increasing the physical size of the system *alone* affects the behavior of the system, particularly as regards the answer to part **(b)**.

T5S.5 (Do Problem T5S.2 first.) Repeat Problem T5S.2 except increase the system's total energy by a factor of 10 (so that $U = 2000\varepsilon$) while keeping the number of atoms in each solid the same ($N_A = N_B = 100$). Compare your answers with those that you found in Problem T5S.2. Comment on how increasing the energy available to the system *alone* affects the behavior of the system, particularly as regards the answer to part **(b)**.

T5S.6 Run the *StatMech* program for two Einstein solids having $N_A = 20$, $N_B = 30$ and $U = 100\varepsilon$. **(a)** On the basis of the principle that the most probable macropartition is the one where the energy per atom in each of the solids is the same, what would you predict the most probable macropartition to be? **(b)** Is your prediction correct? If not, is the actual most probable macropartition at least *close* to your prediction?

T5S.7 Run the *StatMech* program for two Einstein solids having $N_A = 20$, $N_B = 80$ and $U = 100\varepsilon$. **(a)** On the basis of the principle that the most probable macropartition is the one where the energy per atom in each of the solids is the same, what would you predict the most probable macropartition to be? **(b)** Is your prediction correct? If not, is the actual most probable macropartition at least *close* to your prediction?

T5S.8 What is the entropy of an Einstein solid with 5 atoms and an energy of 15ε? Express your answer as a multiple of k_B.

T5S.9 What is the entropy of an Einstein solid with 50 atoms and an energy of 100ε? Express your answer as a multiple of k_B. (*Hint:* you will find that using *StatMech* is by far the fastest way to find the multiplicity here.)

T5S.10 A certain macropartition of two Einstein solids has an entropy of $305.2k_B$. The next macropartition closer to the most probable one has an entropy of $335.5k_B$. If the system is initially in the first macropartition, and we check it again later, how many times more likely is it to have moved to the other than to have stayed in the first?

T5S.11 My calculator cannot display e^x for $x > 230$. Here is a way to calculate e^x for larger values of x. (a) From equation T5.9a, $e^x = 10^{x/\ln 10}$. Note that 10 raised to a non-integer power (for example, $10^{3.46}$) can be broken up as follows: $10^{3.46} = 10^{(3 + 0.46)} = (10^3)(10^{0.46}) = 2.9 \times 10^3$. (b) Use this technique to solve the following problem. The entropy of the most probable macropartition for a certain system of Einstein solids is $6025.3k_B$, while the entropy of an extreme macropartition is only $5755.4k_B$. What is the probability of finding the system at a given time in the extreme macropartition compared to that of finding it in the most probable macropartition?

RICH-CONTEXT

T5R.1 Space aliens deliver into your hand an object made of a substance that can store thermal energy but whose multiplicity actually *decreases* as its energy increases. What will happen to such an object if it is placed in thermal contact with a normal object (like an Einstein solid)? Can this object ever be in thermal equilibrium with a normal object? If you put an object of this substance in a flame, will the flame warm it? How might you increase the thermal energy of an object made of this substance? Do you think that we can assign a meaningful temperature to such an object? Answer these questions *qualitatively*, but carefully, supporting your answers with arguments based on the ideas in this chapter. (*Hint:* What would a macropartition table look like for an object like this in contact with an Einstein solid?)

T5R.2 (Adapted from Kittel and Kroemer, *Thermal Physics*, 2/e, page 53.) The writer Aldous Huxley is reported to have said that "six monkeys, set to strum unintelligently on typewriters for millions of years would be bound in time to write all of the books in the British Museum". This statement is in fact completely *false*: Huxley has been mis-

led by an incorrect intuition about the character of extremely large numbers.

Let's set ourselves a much less difficult problem. Imagine that we have 10 billion monkeys (twice the human population of the Earth) typing diligently at the rate of 2 characters per second since the universe began 5×10^{17} s ago. Say that instead of requiring them to type out an entire library, we'll settle for a single typed version of *Hamlet*. Let's guess that *Hamlet* has approximately 10^5 characters. (a) Assume that the typewriters used by the monkeys have 26 letters and 10 punctuation characters (space, carriage return, period, comma, colon, semicolon, quotation mark, apostrophe, dash, question mark) for a total of 36 characters. We'll ignore the distinction between capital letters and small letters. The probability that any given character is the first character in *Hamlet* is thus 1/36. The probability that this character and the next are the first *two* characters of *Hamlet* is $(1/36)(1/36) = (1/36)^2$. Argue then that the probability of any given sequence of 10^5 random characters being *Hamlet* is $10^{-155,630}$. [*Hint:* $\log_{10}(x^a) = a\log_{10}(x)$, where $\log_{10}$ is the base-ten logarithm. $\log_{10}$ is usually called simply "log" on most calculators.] (b) Argue that the probability that *Hamlet* would be randomly typed by any of our army of monkeys since the beginning of the universe $\approx 10^{-155602}$. [*Hint:* remember that *any* key typed by *any* monkey could in principle be the first character in the play.] This probability is *zero* in any practical sense: even our huge bevy of monkeys will never, never, *never* be able to type *Hamlet* at random.

ADVANCED

T5A.1 Imagine that we can measure the entropy of a system of two solids to within two parts in a billion. This means that we could just barely distinguish a system that has an entropy of 4.99999999 J/K (8 nines!) from one that has 5.00000000 J/K. (a) Imagine that the entropy of the equilibrium macropartition is 5.00000000 J/K. Show that the approximate probability that at any given time later we will find the system in a macropartition with entropy 4.99999999 J/K (that is, with an entropy that is only barely measurably smaller) is about $10^{-31,500,000,000,000}$ times smaller than the probability we will still find it to have entropy 5.00000000 J/K. (*Hint:* see problem T5S.11). (b) Defend the statement that the entropy of an isolated system in thermal equilibrium *never* decreases.

ANSWERS TO EXERCISES

T5X.1 You should find that as N_A, N_B and U get larger, the multiplicities get extremely large and the probability curve becomes more and more sharply peaked.
T5X.2 The probability that if we look at the system we will find it in the 1:179 macropartition is 9.1×10^{-33}.
T5X.3 Roughly 60 chances in a billion (6×10^{-8}).
T5X.4 (Answer is given.)
T5X.5 Having a *huge* number of microstates makes variations in the probabilities between macropartitions extreme, so that some macropartitions have outrageously greater probability than others: this insures the inevitability of movement toward those macropartitions. A sharp peak of probability insures that only a small range of configurations contain the bulk of the microstates, making

substantial fluctuations away from the most probable macropartition unlikely.
T5X.6 The multiplicity of the combined system of two objects in a given macrostate is given by

$$g_{AB} = g_A g_B \qquad \text{(T5.11)}$$

If we take the natural log of both sides, we get

$$\ln g_{AB} = \ln(g_A g_B) = \ln g_A + \ln g_B \qquad \text{(T5.12)}$$

Multiplying both sides of this by k_B and using the definition of S given by equation T5.2, we get equation T5.4.
T5X.7 You should find that the statement is generally true except for *very* small systems. In cases where $U > 200$, the precise most probable macropartition is typically buried in a group, making it tough to be precise.

T6

THE SECOND LAW

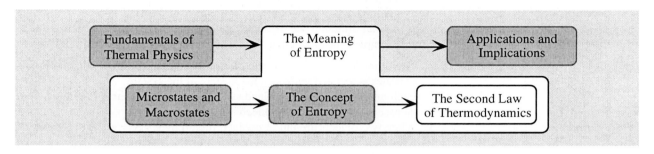

T6.1 OVERVIEW

In the last chapter, we took a giant step towards the understanding of why certain processes are irreversible. In the case of two Einstein solids in thermal contact, we found completely random and reversible microscopic processes (leading to random shuffling between microstates) tend at the macroscopic level to push the solids inexorably toward an equilibrium macropartition, simply because the vast majority of the system's microstates are near to this macropartition.

This is one of the most profound and beautiful ideas in physics, neatly resolving what superficially looks to be an impossible conundrum (how can reversible microscopic processes lead to irreversible macroscopic processes?) with the help of a simple and credible hypothesis (that every accessible microstate of the system is equally probable). We see that it is not so much that a macroscopic process cannot in *principle* go in reverse, it is just that it turns out to be extraordinarily *unlikely* to do so: since the vast majority of microstates available to a multi-object system are close to the equilibrium macropartition, the system inexorably moves toward that macropartition.

Up to this point, we have been exploring this idea in the context of interacting Einstein solids because they are easy to work with. However, one of the beauties of the mechanism for irreversibility discussed in the previous chapter is that it doesn't really depend crucially on the details of the Einstein solid. For reasons that I will discuss in section T6.2, almost any complex system will behave in much the same way. The main purpose of this chapter is to generalize the concept of entropy, discuss the relationship between entropy and "disorder", and explore how entropy is linked to temperature.

Here is an overview of the sections in this chapter.

T6.2 *IRREVERSIBILITY IN GENERAL* discusses what is needed for our explanation of irreversibility to work, and the degree to which common complex systems satisfy these criteria.

T6.3 *THE SECOND LAW OF THERMODYNAMICS* formally defines entropy in general and states the second law of thermodynamics.

T6.4 *ENTROPY AND DISORDER* discusses why entropy is often linked to "disorder" and why this link is partly accurate and partly misleading.

T6.5 *ENTROPY AND TEMPERATURE* explores how we can use entropy to define an object's temperature mathematically (without ever referring to a standard thermometer!).

T6.2   IRREVERSIBILITY IN GENERAL

Consider a hot object of any kind in thermal contact with a cold object of any kind. In the case where the objects are Einstein solids, we have seen that energy will almost inevitably flow from the hot object to the cold object, particularly if the number of atoms and the energy involved is very large. What is really required for the explanation of irreversibility presented in the last chapter to work more generally?

Our explaination of irreversibility is founded on four main ideas

If you look back over the argument presented in Section T5.4, you might be able to see that although we were continually using the *StatMech* program to generate examples, the qualitative argument depends only on four basic ideas:

1. There are a countable number of macroscopically indistinguishable microstates associated with each macrostate of the system,
2. Each microstate of the combined system is equally probable (the fundamental assumption of statistical mechanics),
3. When two objects are put in contact, the product $g_{AB} = g_A g_B$ of their multiplicities is a sharply peaked function of the possible macropartitions of the combined system, and
4. The number of microstates corresponding to a macrostate is *huge*.

A review of why these ideas are important to the argument

Let me briefly describe why each one of these assertions is important. (1) The whole scheme revolves around the basic idea that there are many microstates consistent with a given macrostate, and if these microstates are not countable (at least in *principle*), then we have no way of telling which macropartitions contain more microstates. (2) The fundamental assumption is our only justification for the idea that a macropartition containing more microstates will be more probable than one with fewer microstates. (3) If the multiplicity g_{AB} of a combined system is a very sharply peaked function of macropartition, then it directly implies (a) that when the system is in a macropartition far from the equilibrium macropartition (where g_{AB} peaks), it will tend to evolve in that direction, because the number of microstates available to it sharply increases in that direction, and (b) that once it has reached the equilibrium macropartition, it will tend to *stay* there, because the number of available microstates falls off sharply on either side of the peak. Finally, (4) if the numbers of microstates involved are huge, *tendencies* or *probabilities* become *certainties*, because the probabilities of significant fluctuations away from the equilibrium macropartition (or away from the path to it) become so incredibly tiny that such fluctuations will never occur in any practical sense.

Why these ideas also apply more generally

As I hope I've made clear, we need only these assertions for our explanation of irreversibility. The Einstein solid model and the *StatMech* program simply made the third and fourth items on the list vividly clear. However, do these ideas hold in more general contexts? Let me address each item in turn.

The *first* assertion that a countable set of microstates exists for every macrostate is a simple consequence of quantum mechanics. In quantum mechanics, the energy states of a quanton are quantized and therefore its possible energy states are countable, at least in principle. So as long as quantum mechanics can be used to describe the system (and this is true for every system, as far as we know), this assertion should be valid.

(I should point out in passing that the idea of microstates and macrostates was invented and successfully used for decades before quantum mechanics was invented: early workers in the area simply invented some *ad hoc* rules for counting microstates. The advent of quantum mechanics provided both conceptual and quantitative justification for some rules and correction for others. Indeed, some of the earliest evidence for quantum effects was uncovered when some of the classical ad hoc counting rules failed to coincide with reality in important situations.)

The *fourth* assertion is also valid under a wide range of circumstances, basically because the number of atoms or molecules in any macroscopic object is large (because Avogadro' number is so large) and because the number of "quanta" of energy that get passed around between these atoms or molecules is also large (the quanta of energy is typically so small). Because any macroscopic object contains a huge number of atoms, there are going to be a huge number of different ways to arrange even a few bits of energy among all these atoms. However, because the internal energy of a macroscopic object is fairly large (on the order of many Joules) and the typical difference in atomic or molecular energy levels is so small (tiny fractions of an eV), there are going to be an unimaginably large number of different ways to divide up this energy into tiny bites for each atom.

For most substances, *g* grows very rapidly with *U*

Let me give you a very crude general argument that drives this point home. Imagine that the difference between quantum energy levels (of translational motion, rotation, vibration, or whatever) in the molecules in an object is on the order of magnitude of some energy ε. If the total internal energy of the object is U and the object contains N molecules, then the average amount of energy available to a given molecule will be roughly U/N. This means that the number of possible microstates that are really accessible to a single molecule will then be at least $n = U/N\varepsilon$, since the molecule might have any energy from zero up to (but maybe not much more than) its average share U/N of the energy, and there are $(U/N)/\varepsilon$ possible quantum states for the molecule between these limits. If there are n possible microstates available to one molecule, then there are $n \cdot n = n^2$ microstates available to two molecules, n^3 states to three molecules and so on. Thus, the number of microstates available to N molecules should be *at least*

$$g \approx n^N \approx \left(\frac{U}{N\varepsilon} \right)^N \tag{T6.1}$$

Note that as long as $n = U/N\varepsilon$ is larger than 1 (that is, as long as ε is small enough so that each molecule will typically have a variety of choices of energy level), $g = n^N$ will be a very large number, because raising anything to the 10^{23} power is going to produce a large number. Even $(2)^{10^{23}} \approx 10^{3 \times 10^{22}}$ (which is a 1 followed by 3×10^{22} zeros).

Exercise T6X.1: Use *StatMech* to show that this formula seriously *under-estimates* the multiplicity of Einstein solid macrostates. (If you cannot run *StatMech*, at least check this formula against the data in Figures T5.1 and T5.3.)

This is a very crude calculation, but careful calculations for a wide variety of specific models show that in the case of large N and small ε, multiplicities do tend to end up having the form of $g \approx f(N,...) \cdot (U)^{aN}$, where a is a reasonably small number and $f(N,...)$ is some function of macroscopic variables *other* than U (see Problem T6A.1 to see how this works out for an Einstein solid). The particular formula is really irrelevant, though: my basic point is that for virtually any kind of reasonable physical object, g will be fantastically large.

This crude calculation can help us understand why the combined multiplicity g_{AB} of two macroscopic objects of virtually any kind in thermal contact will be a highly peaked function of configuration. Note that if g is given by anything like equation T6.1, then it increases as U increases (and we hold N and ε fixed) as U^N, which is an *extremely* rapidly increasing function of U.

Exercise T6X.2: Say that $N = 10,000$, which is an infinitesimal number of molecules as macroscopic objects go. Assume $g \propto U^N$, and imagine that the object's initial energy is some value U_i. If the object's energy is increased by 1%, so that $U_f/U_i = 1.01$, by about what factor does the multiplicity increase?

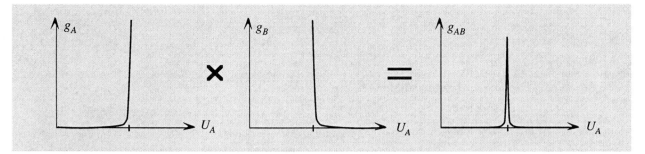

Figure T6.1: The product of rapidly increasing and rapidly decreasing functions is a spike. (In all three graphs, I have defined the vertical axis so that the value of the function at the mark on the horizontal axis is full-scale on the graph's vertical axis.)

_g_ growing rapidly with _U_ ⇒ spike in graph of multiplicity versus macropartition

So g_A for object A is an extremely rapidly increasing function of U_A, almost no matter what kind of object it is. Similarly g_B for object B will be an extremely rapidly increasing function of U_B. When these two objects are placed in thermal contact, though, any growth in U_A comes at the expense of U_B, since the total energy of the combined system $U = U_A + U_B$ is fixed. So as U_A increases, g_A will very rapidly increase, but g_B will very rapidly decrease. The product $g_{AB} = g_A \cdot g_B$ of a rapidly increasing function and a rapidly decreasing function is a very narrow spike, as shown in Figure T6.1.

We also can see this mathematically as follows. Consider two identical N-molecule objects A and B in thermal contact, and let us assume that $g_A \propto U_A^N$ and $g_B \propto U_B^N$. We know that in the *equilibrium* macropartition, these identical objects will each get half of the system's total energy:

$$U_A = U_B = \tfrac{1}{2} U \equiv U_{\text{eq}} \tag{T6.2}$$

Consider now a different macropartition where $U_A = U_{\text{eq}} + x U_{\text{eq}}$ (implying that $U_B = U_{\text{eq}} - x U_{\text{eq}}$) where x is a unitless parameter

$$x \equiv \frac{U_A - U_{\text{eq}}}{U_{\text{eq}}} \tag{T6.3}$$

corresponding to the fractional difference between U_A in this macropartition and its equilibrium value U_{eq}. The multiplicity of this macropartition is

$$g_{AB} = g_A g_B \propto U_A^N U_B^N \propto (1 - x^2)^N \tag{T6.4}$$

Note that this function does indeed have its peak at $x = 0$, where $U_A = U_B = U_{\text{eq}}$.

Exercise T6X.3: Verify equation T6.4. (*Hint:* absorb a factor of U_{eq}^2 into the constant of proportionality.)

Exercise T6X.4: Let's take $N = 100,000$ (again, a fairly small number of molecules as objects go). If $x = \pm 0.01$ (meaning that U_A is different from its equilibrium value by about 1%, what is the value of this g_{AB} compared to its equilibrium value? Is g_{AB} a very sharply peaked function?

Summary

We see then that as long as the multiplicity g for an object is a very rapidly increasing function of its internal energy U, we will have a sharply peaked combined multiplicity function g_{AB} when two objects are placed in contact.

In summary, we see that under quite *general* circumstances,

1. Each macrostate of a macroscopic object has a countable set of microstates,
2. The number of microstates for typical macrostate is fantastically huge, and
3. When two objects are placed in contact, the product of their multiplicities $g_{AB} = g_A g_B$ is a *very* sharply peaked function of the possible macropartitions of the combined system.

To these statements, we only need to add the fundamental assumption of statistical mechanics for the explanation of the irreversibility of heat transfer that I presented in the last chapter to work. Now, there is no way to *prove* the fundamental assumption, or even argue in its favor (as I have done for the other three assertions stated above). It is the founding assumption of the statistical mechanical model, whose validity can only be confirmed or denied on the basis of experiment. Experiments up to the present *do* seem to confirm the qualitative and quantitative predictions of this model.

T6.3 THE SECOND LAW OF THERMODYNAMICS

Let me again summarize why a flow of heat from hot to cold is irreversible according to the model that we have been developing. If we place a hot object in thermal contact with a cold object, energy will flow (on the average) from the hot object to the cold object because there is an enormously greater number of microstates available to the combined system as it moves to a more even distribution of energy as opposed to the opposite direction. Once the system reaches the neighborhood of the most probable macropartition, it will tend to stay there, because the vast majority of microstates available to the system are in macropartitions close to the most probable macropartition.

Up to now, we have been talking exclusively about heat flow, but the fundamental assumption has much broader implications. The macropartition of a system will change in whatever way and by whatever means will take it in the direction of macropartitions having an increasing number of microstates. If exchanging energy between parts of the system is physically possible and increases its total number of microstates, then energy will be exchanged. If exchanging *particles* between parts of the system is physically possible and increases its total number of microstates and such particle flows are possible, then particles will be exchanged. If the system can expand and expansion increases the number of microstates available to the system, then it will expand. Whatever increases the number of microstates will happen if it is allowed by the fundamental laws of physics and whatever constraints we place on the system.

On the other hand, a process taking a system to a macropartition with a smaller number of microstates will never *spontaneously* occur (or more correctly, is fantastically improbable). The macropartitions of a macroscopic system will evolve spontaneously in the direction of increasing multiplicity but never in the direction of decreasing multiplicity.

If we remember that *entropy* is basically another way to talk about *multiplicity*, then this last statement is essentially a statement of the **second law of thermodynamics**: *the entropy of a system never decreases*. With this in mind, let us carefully define entropy in terms more general than we used in the last chapter, and then carefully state a general form of the second law.

A system will evolve to whatever macropartition maximizes its entropy

Definition: The **entropy** of an object whose macrostate is defined by macroscopic variables N, U, ... is defined to be:

$$S = k_B \ln g(N, U, \ldots) \tag{T6.5}$$

where $g(N, U, \ldots)$ is the multiplicity of the macrostate defined by N, U, ...

The general definition of *entropy*

The macrostate of an Einstein solid is completely defined by N and U. The ... in the definition above allows for the possibility that other systems may require more variables to fix the macrostate. To specify the macrostate of an ideal gas, for example, we need three variables: N, U, and V would suffice. Other kinds of systems might require other variables.

Another thing to note is that since the entropy is defined in terms of the macrostate's *multiplicity* (which is a well-defined number for a well-defined macrostate) *the entropy is a function of the macrostate of the system alone.* A system in a given macrostate will have one and only one possible value of *S*.

The general definition of a *macropartition*

Definition: If a system contains two or more interacting subsystems having their own distinct macrostates, the system's *macropartition* is described by specifying the macrostates of the subsystems.

Consequence: If a system contains two or more interacting subsystems having their own distinct macrostates, the total entropy of the system in any given macropartition is the sum of the entropies of the subsystems for the macrostates they have in that macropartition:

$$S_{\text{TOT}} = S_A + S_B + ... \tag{T6.6}$$

Equation T6.6 follows from the fact that we have to *multiply* the multiplicities of systems in contact to find the total multiplicity of the system (that is, $g_{TOT} = g_A \cdot g_B \cdot ...$), the definition of the entropy as the logarithm of the multiplicity, and the fact that $\ln(xy) = \ln x + \ln y$.

The second law

The second law of thermodynamics: The total entropy of an isolated system of interacting objects never decreases.

The *total entropy* of such a system is defined by equation T6.6. It is important to say isolated system, because the entropy of an object *can* decrease if it is interacting with another object. For example, a hot object in contact with a cold object suffers a *decrease* of entropy as it loses energy. But the entropy of the cold object increases more rapidly than the entropy of the hot object decreases, and thus the total entropy of the combined system increases, making this process possible.

The second law is in a certain sense simply a restatement of the fundamental assumption, with the added understanding that because of the huge numbers involved, *probable* become *certain* and *improbable* becomes *impossible* (we have seen how this works with the examples involving Einstein solids in the last chapter).

Note that I have here restated the second law in very general language. Most of what we will be doing in the rest of the book is working out its implications in various situations.

T6.4 ENTROPY AND DISORDER

Many popular treatments link *entropy* to the concept of *disorder*. Here we have defined entropy in terms of *multiplicity*. How are these ideas related?

Examples of disorder linked to multiplicity

Think about it this way. Why does your dorm room get disorderly unless you specifically clean it up? *Because that's where the microstates are.* There are many, many, more ways for your room to be disorderly than orderly. Because of this, random things that happen in your room are *far* more likely to contribute to disorder than to greater order.

Let's consider some more physical examples of disorder. A flowing liquid of a certain kind of atom has more disorder than a neat crystalline solid made of the same atoms. A substance with a sufficient amount of thermal energy will therefore be a liquid instead of a solid because *that's where the microstates are*: there are many, many more microstates available to molecules if they are free to roam around than if they are confined to a solid.

Figure T6.2 shows a container with two compartments. Before mixing, each compartment contains different gases. A valve between the two compartments is then opened, allowing the gases to mix with each other. You may know that under such circumstances, the gases will *spontaneously* intermingle, just like a drop of cream put in a cup of coffee will naturally diffuse throughout the cup (though a little stirring speeds up the process). Why do the gases spontaneously

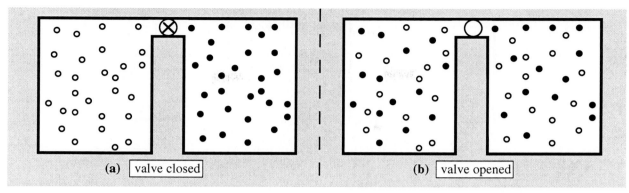

Figure T6.2: Spontaneous mixing of gases.

mix, becoming more "disorderly" in a certain sense? *Because that's where the microstates are.* There are many, many, many more microstates available to the system when the gas molecules are free to roam over the entire container than there are when the gases are separated and confined to their side of the valve.

Figure T6.3 shows N gas molecules constrained by a barrier to be in the left half of the box (the right side contains a vacuum). When the barrier is removed, the gas spontaneously expands to fill the box, becoming more "disorderly" than they were when confined to their proper half (this process is called **free expansion**). Why do they do this? *Because that's where the microstates are.* There are *many* more microstates available to gas molecules free to roam over the entire container than there are for molecules compressed into the left half.

Exercise T6X.5: In fact, we can calculate this. There are 2^N different ways to choose which half of the box each molecule goes into (2 possibilities for each molecule). Only *one* of these ways corresponds to a macropartition where all the molecules are on the left side of the box. So when the barrier is removed, the number of microstates available to the molecules has to increase by a factor of 2^N. If $N = 100$ (a pitifully small number of molecules), by what factor does the multiplicity of the system increase when the barrier is removed?

We see that in many standard situations increased multiplicity is indeed linked with disorder. It is important to remember, however, that entropy is defined in terms of *multiplicity*, not disorder. Increasing entropy does not *necessarily* imply increasing disorder.

In some cases, link with disorder not helpful

For example consider life itself. Life has been described by some people as being anti-entropic, because living things take unorganized inorganic matter and transform it into the most intricate and complex order that one can imagine. Isn't this inconsistent with the law of increasing entropy?

Not at all! Every creature releases heat into its environment as it organizes matter. This heat causes the creature's surroundings to increase in entropy far

Figure T6.3: Free expansion of a gas into a vacuum.

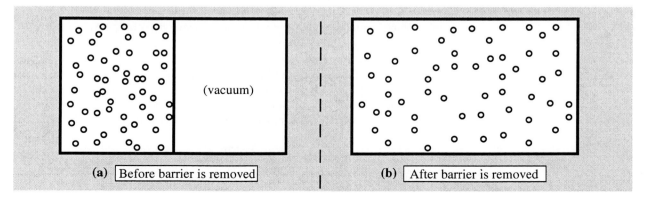

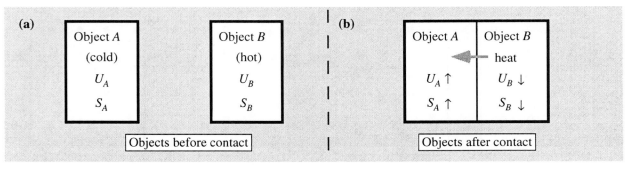

(a)

Object *A* (cold) U_A S_A

Object *B* (hot) U_B S_B

Objects before contact

(b)

Object *A* $U_A \uparrow$ $S_A \uparrow$

Object *B*   heat $U_B \downarrow$ $S_B \downarrow$

Objects after contact

Figure T6.4: When we bring a hot object into contact with a cold object, the entropy of the hot object decreases, while the entropy of the cold object increases.

more than the creature decreases its entropy by putting a few things in order. Indeed, thoughts in a brain, movements of a muscle, growth, development, even evolution itself should be considered to be examples of *entropy in action*. Sources of energy are tapped by living things and ultimately dissipated into the environment, and this natural, spontaneous flow of energy from being concentrated to dissipated is cleverly tapped by organisms to help accomplish a little organization. The beautiful order of life is a manifestation of the dissipation of energy (and consequent increase in entropy of the universe) in much the same way that the orderly ticking of a clock manifests the winding down of its spring. Stars and galaxies form, the sun shines, tectonic plates move, volcanoes erupt, storms form, and spring flowers come because all systems in the universe are evolving toward the macropartitions that maximize their entropy. Every physical process in our daily life, whether associated with growth or decay, is not merely *consistent* with the second law of thermodynamics: it is a *consequence* of that law.

Basic meaning of entropy is multiplicity, not disorder

There are some cases, then, where the idea of *entropy as disorder* can be misleading. Whenever there appears to you to be a contradiction between the concepts of entropy as disorder and entropy as multiplicity, you should remember that *multiplicity* is the more basic.

T6.5 ENTROPY AND TEMPERATURE

Argument linking entropy to temperature

Consider once more our famous *paradigmatic thermal process*, as illustrated in Figure T6.4. A hot object is brought into contact with a cold object. Subsequently heat flows from the hot object to the cold object, decreasing the energy (and thus entropy) of the hot object and increasing the energy (and thus entropy) of the cold object.

Exercise T6X.6: Explain in your own words why if we know that U_A increases, we are almost completely certain that S_A increases with it.

Assuming that the two objects have a large number of molecules, the combined system will evolve inexorably toward the most probable macropartition and subsequently remain there, in *thermal equilibrium*. The thermal equilibrium macropartition will therefore be the macropartition with the greatest number of microstates, and therefore the greatest total entropy $S_{TOT} = S_A + S_B$.

Now, the system's entropy S_{TOT} is a function of the system's macropartition, which in a pure heat transfer process (where V, N, and other macroscopic characteristics of our objects are held constant) is determined by the objects' energies U_A and U_B. Actually, we only need to know U_A to determine the macropartition, since $U_B = U - U_A$, where U is the fixed total energy of the system. The macropartition where S_{TOT} is maximum is specified by the value of U_A such that

$$0 = \frac{dS_{TOT}}{dU_A} \tag{T6.7}$$

(This is the usual way of finding the maximum of a function.) Since the system's total entropy $S_{TOT} = S_A + S_B$, this becomes

$$0 = \frac{d(S_A + S_B)}{dU_A} = \frac{dS_A}{dU_A} + \frac{dS_B}{dU_A} \qquad (T6.8)$$

Now, since whatever thermal energy is gained by A is lost by B, we know that $dU_A = -dU_B$, right? Plugging this into equation T6.8 and rearranging, we get

$$\frac{dS_A}{dU_A} = \frac{dS_B}{dU_B} \qquad \text{(when system is in equilibrium)} \qquad (T6.9)$$

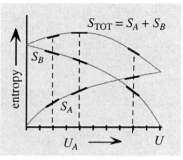

Figure T6.5: The system's total entropy is maximum where the slopes of the entropy curves for parts A and B become equal in magnitude and opposite in sign (at about $U_A = 0.4U$ in the case shown). When this is so, the entropy gained by A when it gets a bit of energy dU_A from B is exactly balanced by the entropy lost by B. If the slopes are not equal the system can gain entropy by transferring energy from one part to the other.

Exercise T6X.7: Check this.

Note that the quantities on either side of equation T6.8 can be calculated *without reference to the other object*: we simply take the derivative of an object's entropy S with respect to its *own* thermal energy U (holding its other macroscopic variables constant). This equation thus tells us that when two objects are in the equilibrium macropartition, the quantity dS/dU calculated for each object must be the same.

Now, the fundamental physical definition of *temperature* (see section T1.5) asserts that it *also* is a quantity that is the same for two objects in equilibrium. Therefore, the quantities dS/dU and temperature T must be related. The most general way that we can write this relationship is to say that for a given object

$$\frac{dS}{dU} = f(T) \qquad (T6.10)$$

where $f(T)$ is some as-yet unknown function of the temperature T.

How can we determine what this function might be? In problem T6A.1, I will argue that the multiplicity of an Einstein solid (in the limit that $3N$ is large and the solid's thermal energy is sufficiently high) is approximately

Determining how dS/dU is related to temperature

$$g(N,U) \approx e^{3N}\left(\frac{U}{3N\varepsilon}\right)^{3N} \qquad (T6.11)$$

To find the entropy, we take the natural logarithm of this multiplicity and multiply by k_B, getting

$$S = 3Nk_B[\ln U - \ln(3N\varepsilon) + 1] \qquad (T6.12)$$

Exercise T6X.8: Check this. Remember that $\ln(x^a) = a\ln x$, $\ln e \equiv 1$, and $\ln(x/y) = \ln x - \ln y$ and $\ln(xy) = \ln x + \ln y$.

Now, if we take the derivative with respect to U (remembering that the derivative of $\ln x$ is $1/x$), we get

$$\frac{dS}{dU} = \frac{d}{dU}\left(3Nk_B[\ln U - \ln(3N\varepsilon) + 1]\right) = \frac{3Nk_B}{U} \qquad (T6.13)$$

We also know from experimental data (see section T4.3) that the thermal energy of elemental solids that seem to fit the Einstein model well is given by

$$U = 3Nk_BT \qquad (T6.14)$$

where T is the ideal gas temperature. Plugging this into equation T6.13, we get

Theoretical definition of temperature

$$\frac{dS}{dU} = \frac{1}{T} \qquad (T6.15)$$

(This equation assumes that other macroscopic variables, such as N, V, and so on, are held constant).

This equation in fact is currently taken to be the theoretical *definition* of absolute temperature, a definition that closely coincides with, but supersedes, the definition of temperature in terms of the constant-volume gas thermometer. This definition is completely independent of any choice of "reference thermoscope": based *only* on the basic principle that temperatures of two objects in contact should be equal when they are in equilibrium and that the definition should at least *approximately* coincide with the temperature scale that we have used before (see problem T6S.5 for some discussion of this issue).

Linking temperature to average energy per molecule

We have also seen in past chapters that temperature is linked to average energy per atom or molecule. Does this follow from our new definition of temperature? In section T6.2, I mentioned that models of a wide variety of thermodynamic systems typically lead to multiplicities of the form

$$g(N, U, \ldots) = f(N, \ldots) \cdot U^{aN} \qquad (T6.16)$$

in the limit of large U and N (note that equation T6.11 can be put in this form, with $a = 3$). For such models, we can easily show that the average energy per molecule in a substance is proportional to T, as the following example shows.

EXAMPLE   T6.1

Problem: Given a substance whose multiplicity is given by equation T6.16, show that the average energy per molecule is proportional to T.

Solution We follow basically the same procedure as we did for the Einstein model earlier in this section. The entropy of the substance is $S = k_B \ln g$, so

$$S = k_B \ln \left[f(N, \ldots) \cdot U^{aN} \right] = k_B \ln f(N, \ldots) + k_B aN \ln U \qquad (T6.17)$$

Taking the derivative of this equation with respect to U, we get:

$$\frac{1}{T} \equiv \frac{dS}{dU} = k_B \frac{d}{dU} \left[\ln f + aN \ln U \right] = 0 + \frac{k_B aN}{U} \qquad (T6.18)$$

since $f(N, \ldots)$ does not depend on U. Multiplying both sides by UT, we get

$$U = aN k_B T \quad \Rightarrow \quad U/N = a k_B T \qquad (T6.19)$$

U/N is the average energy per molecule, so equation T6.19 tells us that any substance whose multiplicity has the form given in equation T6.16 will have an average energy per molecule that is proportional to T.

Equation T6.15 provides an important link between entropy (a rather abstract idea) and temperature and energy (which are more concrete). In the next chapter, we will explore this link in more detail.

SUMMARY

I. IRREVERSIBILITY IN GENERAL
    A. The argument presented in chapter T5 about why heat flow is irreversible ultimately depends on four basic assertions:
        1. Every macrostate has a countable number of microstates
        2. Each accessible microstate is equally probable
        3. When systems get large, multiplicities get outrageously large
        4. The multiplicity is a strongly peaked function of macropartition
    B. The last two were illustrated by a study of Einstein solids in chapter T5

C. But these assertions apply to more general objects as well
   1. The first assertion follows from quantum mechanics
   2. The second is the *fundamental assumption* of statistical mechanics
   3. About the third assertion:
      a) A general (but nonrigorous) argument $\Rightarrow$ $g \approx (U/N\varepsilon)^N$
      b) Actually, *many* models predict that $g \approx f(N,...) \cdot U^{aN}$, where a is some small number and $f(N,...)$ does not depend on U.
      c) This increases fantastically rapidly with U if N is large.
   4. About the fourth assertion: When two objects A, B are in thermal contact, if U_A increases, the $U_B = U - U_A$ must decrease.
      a) Now, g_A goes sharply up and g_B sharply down as U_A increases
      b) The product of sharply increasing and decreasing functions is a spike (a mathematical argument supports this conclusion)

II. THE SECOND LAW OF THERMODYNAMICS
   A. The most general definition of *entropy*: $S = k_B \ln g(N, U, ...)$ (T6.5)
      1. $g(N, U, ...)$ is the multiplicity of the macrostate defined by N, U,...
      2. This means that the entropy is a function of macrostate
   B. A more general definition of *macropartition*: If system contains two or more subsystems, macropartition specified by subsystem macrostates
   C. A system's *total entropy* in a macropartition is $S_{TOT} = S_A + S_B + ...$ (S_A is A's entropy in the macrostate A has in that macropartition, etc.)
   D. *The second law of thermodynamics*
      1. Statement: The *total entropy* of an *isolated* system never decreases
      2. A system's entropy *can* decrease if it is *not* isolated, though

III. ENTROPY AND DISORDER
   A. Examples of spontaneous processes that increase disorder:
      1. random perturbations of your dorm room
      2. mixing of initially separated gases
      3. free expansion of a gas into a vacuum
   B. The basic reason that disorder often increases is because *that's where the microstates are*: many more ways to be disorderly than orderly.
   C. In some cases, "disorder" is not a helpful idea for describing entropy.
      1. Life, spring flowers are examples of entropy in action!
      2. Here, increasing "disorder" (whatever that means) is at best hidden
      3. Fundamentally, entropy is related to *multiplicity*, not disorder

IV. ENTROPY AND TEMPERATURE
   A. Imagine an isolated system of two objects A, B in contact so that they can exchange heat energy (but other macroscopic variables are fixed)
      1. Conservation of energy implies $U_B = U - U_A$
      2. Therefore, specifying U_A completely specifies macropartition
      3. To find the macropartition with max. entropy, find U_A such that

$$0 = \frac{dS_{\text{TOT}}}{dU_A} = \frac{dS_A}{dU_A} + \frac{dS_B}{dU_B}$$

   4. Since $U_B = U - U_A$ we have $dU_B = -dU_A$. Substituting, we get

$$\frac{dS_A}{dU_A} = \frac{dS_B}{dU_B} \quad \text{(in the equilibrium macropartition)}$$

   5. But objects' *temperatures* are equal in equilibrium by zeroth law!
   6. Therefore, $dS/dU = f(T)$, where $f(T)$ is unknown function of T
   7. An Einstein solid argument implies that $dS/dU = 1/T$ (T6.15)
   B. We now take $dS/dU = 1/T$ as the *definition* of temperature
      1. This definition does *not* depend on choice of standard thermoscope
      2. It fortunately coincides with earlier definition of temperature!
   C. Any substance with $g \propto U^{aN}$ has an average energy per atom $\propto ak_BT$

GLOSSARY

entropy S: the entropy of an object in a macrostates specified by macroscopic variables $N, U, \ldots$ is defined to be $S(N, U, \ldots) = k_B \ln g(N, U, \ldots)$, where $g(N, U, \ldots)$ is the multiplicity of the macrostate.

total entropy S_{TOT} (of a macropartition): The macropartition of a system consisting of two or more distinct and interacting objects is specified by stating the macrostates

of each object. The total entropy of the system in that macropartition is the sum of the entropies of each object in their respective macrostates

the second law of thermodynamics: the total entropy of an isolated system never decreases.

free expansion: the irreversible process where a gas expands into previously empty space.

TWO-MINUTE PROBLEMS

T6T.1 The entropy of virtually any object increases as the object's thermal energy increases (T or F).

T6T.2 A system of two or more interacting macroscopic objects is always found in the single most probable macropartition after enough time has passed (T or F).

T6T.3 If N is small for the two objects in thermal contact, the second law of thermodynamics does not particularly accurately apply to this system (T or F).

T6T.4 The multiplicity of an Einstein solid is approximately given by equation T6.11. Consider such a solid with 100 atoms and an initial energy of 600 units. If we increase the solid's energy by 1% to 606 units, the solid's multiplicity increases by about
A.   1%
B.   100%
C.   a factor of 6^{100}
D.   a factor of 6^{300}
E.   a factor of 2^{300}
F.   a factor of $(1.01)^{300}$
T.   other (specify)

T6T.5 When we put a hot object in contact with a cold object, the entropy of one of the two objects actually decreases (T or F).

T6T.6 When water freezes to ice cubes in a freezer, the water's entropy decreases (T or F). This violates the second law of thermodynamics (T or F).

T6T.7 An object whose multiplicity is always 1, no matter what its thermal energy is has a temperature that
A.   is always zero
B.   is always fixed
C.   is always infinite
D.   increases with U
E.   decreases with U
F.   other (specify)

T6T.8 If an object has a multiplicity that *decreases* as its thermal energy increases, its temperature would
A.   be always zero
B.   be always fixed
C.   increase with U
D.   decrease with U
E.   be negative
F.   other (specify)

HOMEWORK PROBLEMS

BASIC SKILLS

T6B.1 Consider Figure T6.1. If g_A is such a rapidly increasing function of U_A, why does it appear to be essentially zero until the mark on the U_A axis? [*Hint*: I defined the vertical scale to be about equal to the value of g_A when U_A is at the mark. When U_A is even just a little less than its value at the mark, what is the value of g_A on this scale?]

T6B.2 Consider an object containing 20,000 molecules. If the object's multiplicity is given by equation T6.1, then by about what factor does its multiplicity increase if the object's energy increases by 1%?

T6B.3 Consider an object containing 5,000 molecules. If the object's multiplicity is given by equation T6.1, by what factor does its multiplicity increase if its energy is increased by 2%?

T6B.4 The second law of thermodynamics states that the entropy of an isolated system never decreases. Your lab partner says that this law is not strictly true, since there is a low probability that the system *could* be found in an extreme macropartition where the multiplicity (and thus entropy) is very small. How would you respond?

T6B.5 Murphy's Law states: "If something can go wrong, it will." Explain how this is really a consequence of the second law of thermodynamics. (*Hint*: How many ways can a given project go wrong compared to the number of ways that it can go right?)

T6B.6 Show that the units are consistent in the equation $dS/dU = 1/T$ that defines temperature. (This is the most important reason for including the factor of k_B in the definition of entropy: it ensures that the temperature defined by this equation is scaled in a manner consistent with the kelvin temperature scale.)

T6B.7 Does it make sense to talk about the temperature of a vacuum? If so, how could you measure or calculate it? If not, why not?

SYNTHETIC

T6S.1 Consider an object containing 2×10^6 molecules. If the object's multiplicity is given by equation T6.1, then by about what factor does its multiplicity increase if its energy increases by 0.1%? (*Hint*: Use logarithms).

T6S.2 When you clean your room, putting everything in order, do you violate the second law of thermodynamics? Explain why or why not.

T6S.3 If you squeeze a gas to a smaller volume while holding its temperature constant, its entropy will decrease, since the system has the same thermal energy as it did before but gas has now less space in which to move and therefore fewer possibilities and thus fewer microstates. Therefore the entropy of such a gas decreases in such a compression. Doesn't this violate the second law of thermodynamics? Explain why or why not.

T6S.4 Consider objects A and B in thermal contact but isolated from everything else. **(a)** Using basically the same argument that was presented near the beginning of section T6.5, show that

$$\frac{dS_{TOT}}{dU_A} = \frac{dS_A}{dU_A} - \frac{dS_B}{dU_B} \tag{T6.20}$$

(b) Using this and the definition of temperature, show that

$$dS_{TOT} = \left(\frac{1}{T_A} - \frac{1}{T_B}\right)dU_A \tag{T6.21}$$

(c) Argue from this that if $T_A > T_B$, energy will spontaneously flow from A to B, but if $T_B > T_A$, then energy will spontaneously flow from B to A, as we in fact observe in nature. [*Hints*: A process will spontaneously occur if it increases the total entropy of a system. If energy flows *into* A, will dU_A be positive or negative?)

T6S.5 In section T6.5, I argued on fairly fundamental grounds that $dS/dU = f(T)$. In principle, we could define $f(T)$ to be anything that we like: this would amount to *defining* temperature and its scale. Still, *some* definitions would violate deeply embedded preconceptions about the nature of temperature. For example, the *simplest* definition of temperature would be $dS/dU \equiv T$. Show that this definition would imply that heat would flow spontaneously from objects with low T to objects with high T. (*Hint*: adapt the result in the previous problem.) This would imply that objects with low T are *hot*, while objects with high T are *cold*. While we *could* define temperature this way, it would really fly in the face of convention (if not intuition).

T6S.6 Imagine that the entropy of a given substance as a function of N and U is given by the formula $S = Nk_B\ln U$. Using the definition of temperature, show that the thermal energy of this substance is related to its temperature by the expression $U = Nk_BT$.

T6S.7 For an ideal monatomic gas, the approximate multiplicity of a system with N atoms and energy U and volume V turns out to be $g(N,U,V) = CV^NU^{3N/2}$, where C is a constant that does not depend on U or V. Use this information and the definition of temperature to determine how the thermal energy of a monatomic gas depends on its temperature.

T6S.8 Imagine that the multiplicity of a certain substance is given by $g(N, U) = Ne^{U/\varepsilon}$ where ε is some unit of energy. How would the temperature of an object made out of this substance depend on its energy? Does this seem realistic?

T6S.9 Imagine that the multiplicity of a certain substance is given by $g(N,U) = e^{-N^2\varepsilon/U}$ where ε is some unit of energy. How would the thermal energy of an object made out of this substance depend on its temperature?

T6S.10 Imagine that the combined multiplicity of a system of two identical objects A and B in thermal contact is given by $g_{AB} \propto (1-x^2)^N$, where x is the unitless parameter defined by equation T6.3. Let us say that the largest fluctuations that we are ever likely to see correspond to those macropartitions where g_{AB} is greater than 10^{-4} of its value at equilibrium (macropartitions at the extreme edges of this range are thus 10,000 times less likely to be seen than

macropartitions near the center). **(a)** Argue that if N is 100, the range for x satisfying this criterion is $|x| < 0.30$, meaning that U_A might vary over a range equal to about ±30% of its equilibrium value. **(b)** Argue that if N = 10,000, this range is $|x| < 0.03$.

T6S.11 Consider the same situation described in the previous problem. **(a)** If $a \ll 1$, it turns out that $\ln(1+a) \approx a$ (as you can check with your calculator). Use this to show that for general values of N, the range of values for x corresponding to macropartitions we are ever likely to see (according to the criterion in the previous problem) is

$$|x| < \sqrt{\frac{\ln(10^4)}{N}} \tag{T6.22}$$

(b) If the two objects in contact contain roughly a mole of atoms each, about how large will be the likely fluctuations in U_A compared to its equilibrium value?

T6S.12 Consider an Einstein solid having N = 20 atoms. **(a)** What is the solid's temperature when it has an energy of 10ε, assuming that $\varepsilon = \hbar\omega \approx 0.02$ eV? Calculate this directly from the definition of temperature by finding S at 10ε and 11ε, computing $dS/dU \approx [S(11\varepsilon)-S(10\varepsilon)]/e$, and then applying the definition of temperature. (You will find that your work will go faster if you use *StatMech* to tabulate the multiplicities.) **(b)** How does this compare with the result from the formula $U = 3Nk_BT$ (which is only accurate if N is large and $U/3N\varepsilon \gg 1$)? **(c)** If you have access to *StatMech*, repeat for N = 20.

RICH-CONTEXT

T6R.1 Try this home experiment. Take one end of a rubber band in each hand and hold the slack rubber band against your lips (which are very sensitive to changes in temperature). Then suddenly stretch the rubber band. You should notice that it gets noticeably warmer. Why? Consider the fact that when the rubber band is slack, the rubber molecules are all tangled up in a complicated way, but when you stretch them out, they become more linear (and thus have fewer ways available to them to be tangled). If the entropy of the rubber increases during the stretching process, some of the energy that you supply while you stretch the rubber band must go to increasing the thermal energy of the rubber. Explain why.

T6R.2 Imagine that space aliens deliver into your hands two identical objects made of substances whose multiplicity increases linearly with thermal energy, something like $g = aNU/\varepsilon$, where ε is some energy unit and a is some constant. Answer the following questions about these objects. **(a)** Do they have a well-defined temperature? If so, how does this temperature depend on the objects' thermal energy? **(b)** If these objects are placed in thermal contact, will energy spontaneously flow from hot to cold? Will the objects eventually come into equilibrium at a certain common temperature? [*Hint*: I suggest drawing a graph of g_{AB} versus macropartition. This will also help in the next part.] **(c)** How will the size of random fluctuations in the energies of these objects compare to those for two normal objects placed in thermal contact? **(d)** In what other ways (if any) will these objects behave differently than normal objects.

ADVANCED

T6A.1 The expression for the multiplicity of an Einstein solid that we found in Chapter T4, when expressed in terms of U and N is:

$$g(N,U) = \frac{(3N + U/\varepsilon - 1)!}{(3N-1)!(U/\varepsilon)!} \qquad (T6.23)$$

where ε is the energy difference between energy levels in the individual quantum oscillators. When N is large, we can ignore the 1 in comparison to $3N$, so this becomes:

$$g(N,U) \approx \frac{(3N + U/\varepsilon)!}{(3N)!(U/\varepsilon)!} \qquad (T6.24)$$

Now, there is an interesting approximation (the *Stirling approximation*) for the logarithm of $m!$ when m is large:

$$\ln(m!) = m \ln m - m \quad \text{when } m \gg 1 \qquad (T6.25)$$

(a) Using this approximation and *assuming* that the ratio $U/3N\varepsilon \gg 1$ (that is, that the object's total energy is large enough so there is plenty of energy to give each oscillator many ε's worth of energy), and using the approximation $\ln(1+x) \approx x$ when $x \ll 1$, show that $\ln g$ is approximately

$$\ln g = 3N \ln\left(\frac{U}{3N\varepsilon}\right) + 3N \qquad (T6.26)$$

(b) Argue that if we take the exponential of both sides of this equation, we get

$$g \approx e^{3N}\left(\frac{U}{3N\varepsilon}\right)^{3N} \qquad (T6.27)$$

as stated in Equation T6.11 near the end of Section T6.5.

T6A.2 Consider a single quanton (such as an atom or molecule) in thermal contact with a reservoir at temperature T. The purpose of this problem is to answer the following question: what is the probability of finding the quanton to be in a certain specific quantum state of energy E relative to its probability of being in its lowest quantum state (which we will define to have zero energy)?

Consider the combined system of the quanton and reservoir. The probability of any macrostate of this combined system is proportional to the multiplicity of that macropartition, which is the *product* of the multiplicities of the quanton and the reservoir. Therefore

$$\text{prob}(E) \propto g_A(1, E) \cdot g_B(N, U-E) \qquad (T6.28)$$

where $\text{prob}(E)$ is meant to represent the probability that the quanton has energy E.

However, the multiplicity for the quanton to be in a given specific quantum state is 1 by definition (there is only one way for the quanton to be in that state). So

$$\text{prob}(E) \propto g_B(N, U-E) \qquad (T6.29)$$

(a) Use the definition of entropy and properties of the exponential function to argue that the probability of the quanton being in a state with energy E as a fraction of the probability that it is in the ground state with zero energy is

$$\frac{\text{prob}(E)}{\text{prob}(0)} = \exp([S_B(U-E) - S_B(U)]/k_B) \qquad (T6.30)$$

where $\exp(x) \equiv e^x$, and S_B is the entropy of the reservoir, expressed as a function of the reservoir's energy.
(b) Note that the energy E is likely to be infinitesimal compared to U. Use this to argue that the quantity in square brackets above is the same as $-E(dS_B/dU)$.
(c) Use the definition of temperature to argue that

$$\frac{\text{prob}(E)}{\text{prob}(0)} = \exp(-E/k_B T) \qquad (T6.31)$$

The factor $\exp(-E/k_B T)$ is called the **Boltzmann factor**. This factor appears in a number of equations when we consider the thermal behavior of microscopic systems.
(d) How many times *less* likely is a hydrogen molecule in a room at room temperature to be in its first excited vibrational energy state ($E = 0.27$ eV) as opposed to its ground state ($E = 0$). Can you see why the vibrational mode of energy storage for this molecule is effectively "frozen out" at room temperature, as discussed in Chapter T2?

ANSWERS TO EXERCISES

T6X.1 When $U = 6\varepsilon$ and $N_B = 1$ (as in Figure T5.1), then $(U/N\varepsilon)^N = (6)^1 = 6$, which is much smaller than the actual multiplicity $g_B = 28$ that the solid has in that macrostate according to *StatMech*. Similarly, when $U = 599\varepsilon$ and $N_B = 100$, $(U/N\varepsilon)^N = (5.99)^{100} \approx 5.5 \times 10^{77}$, which is *much* smaller than the multiplicities of order 10^{246} shown in Figure T5.3. Other cases you might try will be similar. Thus any argument that we make about how large the numbers produced by equation T6.1 are, the argument is even *more* true for Einstein solids.
T6X.2 By about a factor of about 1.6×10^{43}.
T6X.3 Note that $U_A = U_{eq}(1+x)$ and $U_B = U_{eq}(1-x)$. Therefore $g_{AB} = g_A g_B \propto U_A^N U_B^N = U_{eq}^{2N}(1+x)^N(1-x)^N = U_{eq}^{2N}[(1+x)(1-x)]^N = U_{eq}^{2N}(1-x^2)^N$. If we absorb U_{eq}^{2N} into the constant of proportionality, we get equation T6.4.
T6X.4 g_{AB} goes down by a factor of about 22,000.
T6X.5 By a factor of about 1.3×10^{30}.

T6X.6 For virtually any kind of substance we can imagine, the more units of energy an object has, the more ways that we can arrange that energy among its objects. In general, g is in fact a very rapidly increasing function of U, so $S = k_B \ln g$ should increase with U as well.
T6X.7 Since $dU_A = -dU_B$, $dS_B/dU_A = dS_B/(-dU_B) = -dS_B/dU_B$. Plugging this into equation T6.8 and then adding dS_B/dU_B to both sides, we get equation T6.9.
T6X.8 We have

$$\ln g = \ln\left[e^{3N}\left(\frac{U}{3N\varepsilon}\right)^{3N}\right] = \ln[e^{3N}] + \ln\left[\left(\frac{U}{3N\varepsilon}\right)^{3N}\right]$$

$$= 3N \ln e + 3N \ln\left(\frac{U}{3N\varepsilon}\right)$$

$$= 3N(1) + 3N[\ln U - \ln(3N\varepsilon)] \qquad (T6.32)$$

So $S = k_B \ln g = k_B 3N[1 + \ln U - \ln(3N\varepsilon)]$.

CALCULATING ENTROPY CHANGES

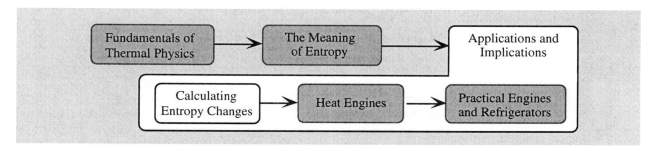

T7.1 OVERVIEW

In the last three chapters, we have been working to develop our understanding of the physics of irreversibility. This process culminated in the last chapter with a broad definition of the concept of entropy, a general statement of the second law of thermodynamics, and a definition of temperature in terms of entropy.

In the next three chapters, we will be exploring applications and implications of the second law, particularly with regard to *heat engines*, devices such as automobile engines and steam engines, that convert heat energy to mechanical energy. In this chapter, we will prepare for our exploration of heat engines by learning how to calculate the change in entropy of a system without having to have a model for the system that allows us to calculate the number of microstates. In the next chapter (chapter T8), we will apply this knowledge to understand the stringent limits that the Second Law places on *all* heat engines, no matter how clever their design. Finally, in chapter T9, we will draw on the concepts we developed in chapter T3 to explore some practical designs for heat engines and refrigerators (which turn out to be closely related).

Here is an overview of the sections in this chapter.

T7.2 *ENTROPY AND TEMPERATURE* reviews the definition of temperature presented in the last chapter and discusses some of its subtleties.

T7.3 *ADIABATIC VOLUME CHANGES* argues that the entropy of a gas undergoing a slow volume change is *constant*, in *spite* of the fact that the thermal energy U of the gas changes.

T7.4 *CALCULATING ENTROPY CHANGES* discusses the general procedure for calculating entropy changes in light of the previous section.

T7.5 *CONSTANT-TEMPERATURE PROCESSES* discusses how to apply this procedure in cases where the temperature of the object is constant.

T7.6 *HANDLING CHANGING TEMPERATURES* explores applying this procedure in processes where the temperature is *not* constant.

T7.7 *NON-QUASISTATIC PROCESSES* presents a clever procedure for calculating entropy changes even in very complicated processes.

Understanding the limitations that the Second Law places on heat engines will involve computing entropy changes using the methods and concepts of this chapter, so working through this chapter will provide a firm foundation for our work in chapter T8.

T7.2   ENTROPY AND TEMPERATURE

A review of the definition
of temperature

In section T6.5 of the previous chapter, we developed a wholly theoretical definition of absolute temperature. If an object in a certain macrostate has entropy S, then the object's temperature in that macrostate is defined in terms of the rate at which the entropy changes with thermal energy

$$\frac{1}{T} \equiv \frac{dS}{dU} \qquad\qquad (T7.1a)$$

The argument in that section was based on three crucial principles: (1) that the total entropy S_{TOT} of an isolated system consisting of two interacting objects is maximum when the objects are in equilibrium, (2) that the total energy of the system is conserved, and (3) that the most basic property of temperature (as expressed by the zeroth law of thermodynamics) is that temperatures of objects in equilibrium must be equal. These principles suffice to establish that dS/dU for an object must be some function $f(T)$ of the temperature. We *chose* this function to be $1/T$ so that the temperature defined by equation T7.1 would coincide at least approximately with the constant-volume gas temperature scale that we had been using previously.

While we made this choice for the sake of historical continuity, it is important to understand that equation T7.1 is in fact a *definition* of temperature T based entirely on its most fundamental characteristic of "equality in equilibrium". Even if we knew *nothing* about the constant-volume ideal gas thermoscope, we might well have chosen this definition anyway, because (1) $dS/dU = f(T) = 1/T$ is the *simplest* definition of temperature consistent with the principle that hot objects should have numerically higher T than cold objects (the only simpler definition $dS/dU = T$ is not consistent with this principle: see problems T6S.4 and T6S.5 in the previous chapter), and because (2) the choice $f(T) = 1/T$ ends up implying that an object's thermal energy per molecule is typically proportional to its temperature, which is convenient. Equation T7.1 therefore is a more *fundamental* definition of temperature than our previous definition based on the ideal gas, and supersedes it in any case where they disagree.

This formula will be the basis for everything that we are going to do in this chapter. In much of what follows, we will actually use the formula backwards: instead of using known values of dS and dU for an object to compute its temperature, we will use known values of T and dU during a process to compute the entropy change dS during that process, allowing us to find entropy changes even when we do *not* have a good model of a substance that would allow us to compute multiplicities directly.

**This definition is limited to
processes with N, V, etc.
held constant**

Before we get into this process, though, we need to understand some of the fine print associated with this definition. We derived equation T7.1 in section T6.5 assuming that the entropy of each of our interacting objects in the situation at hand depended only on its internal energy U. However, we know that even for an object as simple as an Einstein solid, the entropy also depends strongly on the number of molecules N in the object. The entropy of a gas also depends strongly on its volume V: the larger V is, the greater the number of microstates available to the gas and thus the greater its entropy (see section T6.4). Therefore, the entropy of two objects will depend only on U only in a situation where the object's values of N and V are fixed. Therefore, technically, we should write the definition of temperature as follows

$$\frac{1}{T} \equiv \left[\frac{dS}{dU}\right]_{N,V,\dots = constant} \equiv \frac{\partial S}{\partial U} \qquad\qquad (T7.1b)$$

where the "curly ∂s" indicate (as in Unit E) a partial derivative where all other variables that S depends on are treated as constants when we take the derivative.

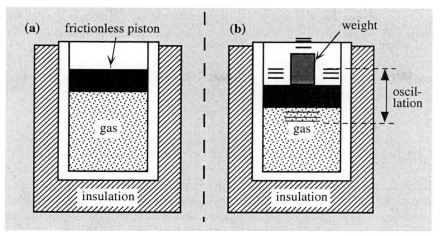

Figure T7.1: (a) A frictionless piston holds a gas confined in an insulated cylinder. **(b)** If the piston is set oscillating, it will continue to do so indefinitely.

In fact, for most solids and many liquids, V is pretty much determined by N and we usually consider situations where objects have a fixed number of molecules N, so we don't have to worry much about the fine print when applying equation T7.1 to solids and liquids. The problem comes when we try to apply the equation to gases, where in certain kinds of processes, changes in volume can be significant. For gas processes, we need a slightly different version of equation T7.1 that we will describe in section T7.4.

T7.3 ADIABATIC VOLUME CHANGES

The key to developing this new version of equation T7.1 is to understand that if a gas changes volume **quasistatically** (slowly enough so that the gas is almost in equilibrium at every point during the volume change) and **adiabatically** (so that no heat flows into or out of the gas) then its change in entropy during the process turns out to be *zero*:

A basic claim about ΔS in adiabatic volume changes

$$\Delta S = 0 \quad \text{(for an adiabatic, quasistatic volume change)} \qquad \text{(T7.2)}$$

How do we know this? There are several ways to argue that this must be true. In this section, I will focus on presenting a physical and more or less intuitive argument. If you'd like to see a mathematical argument as well, I provide one in the context of problem T7A.1.

Consider the situation shown in Figure T7.1. An ideal gas is placed in an insulated cylinder capped with a fairly massive but frictionless piston that is initially at rest. Then imagine that I put an extra weight on the piston. The piston is now too heavy to be supported by the gas pressure, and so it begins to accelerate downward. As the piston moves downward, the gas' volume decreases and thus its pressure increases according to the equation

An argument for the claim

$$PV^\gamma = \text{constant} \qquad \text{(T7.3)}$$

The increasing force exerted by this pressure on the piston eventually matches the piston's weight, but by this time, the piston has downward momentum, so the piston overshoots its equilibrium position, compressing the gas further. The increased gas pressure eventually stops the piston and begins to accelerate it the other way, where it overshoots the equilibrium position, finally coming to rest at the position it had when the weight was added. However, just as before, the gas pressure is too small to support the piston so it begins to move downward again and the cycle repeats. The piston will thus oscillate as if the gas inside the cylinder were a spring, like a child bouncing up and down on an inner tube.

In order for the next part of my argument to make sense, you have to believe that if the piston is truly frictionless, the gas is completely ideal, and the bouncing is sufficiently slow, this oscillation will go on indefinitely. You may be able to see this intuitively from your experience. If you are still unsure, let me point out the following. The pressure of the gas (and thus the force it exerts on the piston) is a function *only* of the volume of the gas according to the expression PV^{γ} = constant. So at any given position of the piston during its motion, the pressure (and thus its effect on the piston's motion) will be the same whether it is the first oscillation or the tenth or the hundredth. The gas *itself*, therefore, will not cause the piston's oscillatory behavior to change.

If it is really true that the piston would ideally oscillate forever, I claim that the oscillation *cannot* be increasing the entropy of whatever isolated system is involved here. A movie of the oscillating piston run forward or backward would look the same, so the process is reversible. Since the second law implies that entropy cannot decrease, the entropy of a reversible process must be constant.

This does not necessarily imply that the *gas'* entropy is constant. Perhaps its entropy *increases* during the compression (for example). As long as some other object's entropy *decreases* at the same time (so that the system's *total* entropy remains fixed), the whole process could still be reversible.

In this case, however, there is nothing else whose entropy could change! The insulation thermally isolates the gas from its surroundings. While the gas does physically interact with the piston, simple changes in piston's **mechanical energy** (its macroscopic kinetic energy and potential energy of external interactions) do not change its entropy: entropy has to do with possibilities associated with rearranging energy storage at the *microscopic* level (that is, with its *thermal energy*), not with the piston's motion as a unit.

Therefore, if the gas does not change the entropy of the things with which it interacts, and yet the whole system's entropy does not change, it follows that the entropy of the gas itself does not change, neither during the compression nor the expansion phase of the oscillation. Therefore, *the entropy of an ideal gas during a quasistatic adiabatic volume change does not change.*

Of course, this oscillation *will* die out in realistic situations because friction in the piston will convert the energy of motion to thermal energy in the piston and the cylinder walls, but this has nothing to do with the gas. A small amount of the energy of motion might go to producing sound waves that randomly bounce around in the gas until their energy is eventually converted into thermal energy. If the process is done very slowly, though, hardly any sound waves will be produced and the *gas* will not contribute to the irreversibility of the process (this is part of why it is important that the process be done *quasistatically*).

Exercise T7X.1: Let us think about the difference between an *adiabatic* and *isothermal* volume change in this argument. Imagine that instead of insulating the cylinder, we immerse it in a bath that keeps it at a constant temperature. Assume for the sake of argument that the piston in this case oscillates also (in practice, we would find that it will oscillate for many fewer cycles than the adiabatic case). Pinpoint the place in the argument above that would be different, and explain why in the isothermal case, even if the piston does oscillate forever, we cannot be *sure* that the entropy change of the *gas* remains zero.

T7.4 CALCULATING ENTROPY CHANGES

Thermal energy *U* changes in an adiabatic process

Note that even though the *entropy* of the gas remains constant in this oscillation process, its thermal *energy* does not. As the piston compresses the gas adiabatically, work energy flows into the gas as the piston's kinetic energy is converted to thermal energy. Similarly as the gas expands adiabatically, work flows *out* of the gas and is converted to kinetic energy in the piston. Therefore, the thermal energy of the gas *does* change during an adiabatic volume change.

Normally, if we increase the internal energy of an object, its entropy will increase. What is going on here is that as the gas is adiabatically compressed, the increase in its entropy arising from the addition of thermal energy is precisely *canceled* by the decrease in entropy associated with the gas being confined to a smaller volume, and the reverse during expansion (see problem T7A.1 for the details). We see that in this case, then, *thermal energy added to the gas by quasistatic expansion or compression has no effect on the gas' entropy.*

So energy added by quasi-static work doesn't count

This provides the foundation that enables us to calculate entropy changes for other gas processes involving volume changes. We will assume that we can treat any such process as an adiabatic volume change plus an additional transfer of thermal energy in the form of heat or other kinds of work. Since we know that the adiabatic part of the process does not change the gas' entropy, it follows that if the gas' entropy *does* change in a process, then, it must be due to thermal energy added to the gas in ways other than quasistatic volume changes.

If there were *no* change in V or N, the definition of temperature implies that

$$\frac{1}{T} \equiv \frac{dS}{dU} \quad \Rightarrow \quad dS = \frac{dU}{T} \quad (N \text{ and } V \text{ constant}) \tag{T7.4}$$

If we allow the gas to change volume quasistatically, then we have to subtract the part of dU associated with the work energy transferred during the quasistatic volume change: this thermal energy does not count. Therefore

$$dS = \frac{dU - dW_{qs}}{T} \quad (N \text{ constant, } V \text{ changes quasistatically}) \tag{T7.5}$$

where dW_{qs} is the work associated with the quasistatic volume change. Under many circumstances, this is the *only* kind of work involved. If so, then the only other way that energy can be added to the gas is in the form of heat, so

$$dS = \frac{dQ}{T} \quad (N \text{ fixed, } V \text{ changes quasistatically, no other work}) \tag{T7.6}$$

A basic formula to use to compute entropy changes

We will spend the remainder of the chapter exploring the uses of equation T7.6.

We derived equation T7.6 assuming that our object was an ideal gas. In practice, however, we can use equation T7.6 for solids and liquids as well. Since most solids and liquids change volume only slightly if at all during typical processes, we can treat V as fixed. If N is also fixed, then equation T7.4 applies. However, as long as no *other* kind of work energy enters the solid or liquid in question, then $dU = dQ$, meaning that equations T7.4 and T7.6 are equivalent.

Equation T7.6 does have its limitations: there are processes to which it does not apply. When a process involves energy transfers to thermal energy that are not heat and not due to quasistatic volume changes (for example, when gasoline explodes in an automobile engine cylinder), equation T7.6 does not apply. Equation T7.6 also does not cover sudden or violent volume changes.

This formula does not apply in *all* cases

Surprisingly, it turns out that with a certain amount of cleverness, we can get around these limitations and apply equation T7.6 indirectly to such processes, even though it does not strictly apply. We'll see how in section T7.7.

Exercise T7X.2 Which of the following processes are consistent with the limitations on equation T7.6 and which are not?
a.   A stone is thrown into a pond.   b.   Soup is slowly heated on a stove.
c.   A cup of coffee is gently stirred.   d.   A bubble slowly rises in a lake.
e.   A gas is slowly compressed in a cylinder while its temperature is held fixed.

T7.5   CONSTANT-TEMPERATURE PROCESSES

Equation T7.6 strictly applies only to heat transfers that are "sufficiently small" so that the temperature doesn't change significantly during the process. In general as we transfer heat to an object, its temperature will change. We cannot calculate ΔS for a finite heat transfer ΔQ unless we take account of the changing temperature by doing an integral:

$$dS = \frac{dQ}{T} \quad \Rightarrow \quad \Delta S = \int_{\text{process}} \frac{dQ}{T} \qquad \left(\begin{array}{l} \text{quasistatic volume change} \\ \text{no other work done, } N \text{ fixed} \end{array} \right) \qquad \text{(T7.7)}$$

We will discuss the procedure for evaluating such integrals in the next section.

If, on the other hand, the temperature of an object remains approximately constant during the process, then whatever amount of heat is transferred during the process *is* "sufficiently small" and we can apply equation T7.6 directly:

The entropy change when temperature is ≈ constant

$$\Delta S = \frac{\Delta Q}{T} \qquad \text{(if } T \approx \text{ constant during a process)} \qquad \text{(T7.8)}$$

Under what circumstances will T be approximately constant? In chapter C9 of unit C (and again at the end of chapter T1) we discussed how the change dU in object's thermal energy as its temperature changes by dT is given by

$$dU = mcdT \qquad \text{(T7.9)}$$

where m is the object's mass and c is its specific heat* (note that specific heats* for various substances are listed on the inside front cover of this text). Imagine an object so massive that it can absorb or supply a significant amount of heat dU while suffering only the tiniest change in temperature dT. The technical term for such an object in thermal physics is a **reservoir**. Thus, we can use T7.8 to:

Three practical situations where $T \approx$ constant

1. compute the entropy change of a *reservoir* absorbing or supplying heat, or

2. compute the entropy change of something in thermal contact with a reservoir, and thus whose temperature is the same as that of the reservoir.

In both cases, the presence of the reservoir ensures that the temperature T appearing in equation T7.8 does not change significantly during the process.

There is a third (unrelated) case where we can use equation T7.8. We saw in unit C that when a substance undergoes a **phase change** (for example, from a solid to a liquid), its thermal energy changes as follows

$$\begin{aligned} \Delta U &= -mL \quad \text{(gas to liquid, liquid to solid)} \\ &= +mL \quad \text{(solid to liquid, liquid to gas)} \end{aligned} \qquad \text{(T7.10)}$$

where m is the mass of the substance and L is its **latent heat*** (latent heats* for various substances are also listed on the inside front cover). The other interesting feature of phase changes is that the temperature of the substance remains *constant* until the phase change is complete. Thus we can use equation T7.8 to

3. compute the entropy change of a substance during a phase change.

These three cases, then, are the most common practical situations where we can apply equation T7.8. The following examples illustrate its use.

EXAMPLE T7.1

Problem: Imagine that we have a bathtub full of water at 20°C and we place a 1.0-kg stone in it whose original temperature is 95°C. What is the entropy change of the water in this case?

Solution The tub of water probably contains hundreds if not thousands of kilograms of water, and its temperature change in this process will be tiny compared to that of the stone. This means that the final equilibrium temperature of both will be about 20°C. According to the inside front cover, the specific heat* of granite is about 760 J·kg^{-1}·K^{-1}, so the heat energy ΔQ_s lost by the stone is

$$\Delta Q_s = m_s c_s \Delta T_s = (1.0 \text{ kg})(760 \text{ J·kg}^{-1}\text{·K}^{-1})(-75\text{ K}) = -57,000 \text{ J} \qquad (T7.11)$$

The water gains this energy, so the water's change in entropy is

$$\Delta S = \frac{\Delta Q_w}{T_w} = \frac{+|\Delta Q_s|}{T_w} = \frac{+57,000 \text{ J}}{293 \text{ K}} = +195 \text{ J/K} \qquad (T7.12)$$

This is positive, as we would expect for something gaining thermal energy.

EXAMPLE T7.2

Problem: Imagine that we have a cylinder containing an ideal gas in good thermal contact with a reservoir at 32°C. Imagine that we slowly compress the gas, doing 45 J of work on it while its temperature remains constant. What is the change in entropy of the gas? What is the change in entropy of the reservoir?

Solution Since the gas' temperature does not change, its thermal energy does not change. Therefore any work energy that it gains in this compression must flow out to the reservoir in the form of heat. The gas therefore *loses* 45 J of heat in this process, so its entropy change (noting that $T = [273 + 32] \text{ K} = 305 \text{ K}$) is

$$\Delta S_g = \frac{\Delta Q_g}{T_g} = \frac{-45 \text{ J}}{305 \text{ K}} = -0.15 \text{ J/K} \qquad (T7.13)$$

(Remember that the work energy it gains in the "slow" compression doesn't count!). The 45 J the gas loses flows *into* the reservoir, so its entropy change is

$$\Delta S_R = \frac{\Delta Q_R}{T_R} = \frac{+45 \text{ J}}{305 \text{ K}} = +0.15 \text{ J/K} \qquad (T7.14)$$

Note that the *net* entropy change of the system in this process is zero.

EXAMPLE T7.3

Problem: Imagine that 120 g of ice at 0°C melts to a puddle of water at 0°C on a surface in a room where the temperature is 28°C. What is the change in the entropy of the water?

Solution According to the table on the inside front cover, the latent heat* associated with the transformation of solid to liquid water is 333 kJ/kg, so the total energy that the ice must absorb from the warm room is

$$dQ = dU = +mL = (0.12 \text{ kg})(333 \text{ kJ/kg}) = 40 \text{ kJ} \qquad (T7.15)$$

Its entropy change is therefore

$$\Delta S_w = \frac{\Delta Q_w}{T_w} = \frac{+40,000 \text{ J}}{273 \text{ K}} = +150 \text{ J/K} \qquad (T7.16)$$

Note that the temperature of the room is irrelevant here.

Exercise T7X.3: If the table surface in the last example acts like a reservoir, find its entropy change. Does the *total* entropy of water and table increase?

Exercise T7X.4: Note that in each of these three examples, I carefully converted temperatures to *kelvins* before computing the entropy. Further, note that if I had divided by temperatures in °C, I would have gotten *very* different answers. Why must we use kelvin temperatures when we use equation T7.8?

T7.6 HANDLING CHANGING TEMPERATURES

Now let us consider cases where we would like to compute the entropy change of an object as it undergoes a heat transfer that changes its temperature significantly. Assume the object has a specific heat* c and mass m. If the object's volume doesn't change (much) during the process, then no significant work is done and:

$$dQ \approx dU = mc \, dT \qquad \text{(if } no \text{ work is done)} \qquad (T7.17)$$

Plugging this into equation T7.7, we get:

$$\Delta S = \int_{\text{process}} \frac{dQ}{T} = \int_{\text{process}} \frac{mc \, dT}{T} \qquad \text{(if no work is done)} \qquad (T7.18)$$

Note that if we consider c to be a function of temperature, this is simply an integral over some function of the object's changing temperature.

If the object's specific heat* is approximately independent of temperature (as it often is over reasonably small temperature ranges), then we can pull both m and c out in front of the integral. Integrating the dT/T that remains from the object's initial temperature T_i to its final temperature T_f, we get

The change in entropy for a heat exchange with no work involved

$$\Delta S = mc \ln\left(\frac{T_f}{T_i}\right) \qquad \text{(if no work is done and } c \approx \text{ constant)} \qquad (T7.19)$$

Exercise T7X.5: Verify equation T7.19.

EXAMPLE T7.4

Problem: Imagine that I place an aluminum block with mass $m = 260$ g and an initial temperature $T_b = 89°C$ into a cup of water with mass $M = 320$ g and an initial temperature $T_w = 22°C$, and allow the two to come to thermal equilibrium. What is the entropy change of the water in this process?

Solution The first step is to find the final equilibrium temperature T_f. The energy flowing out of the metal goes into the water, so $dQ_w, = -dQ_b$ implying that

$$MC(T_f - T_w) = dQ_w = -dQ_b = -mc(T_f - T_b) \qquad (T7.20)$$

where c and C are the specific heats* of the block and water respectively, and dQ_b and dQ_w are the amounts of heat gained by each respectively. Solving for T_f, we get

$$(MC + mc)T_f = MCT_w + mcT_b \quad \Rightarrow \quad T_f = \frac{MCT_w + mcT_b}{MC + mc} \qquad (T7.21)$$

Using $C = 4186$ J·kg⁻¹·K⁻¹ and $c = 900$ J·kg⁻¹·K⁻¹ (see inside front cover), and converting $T_w = 22°C$ to (273+22) K = 295 K (similarly, $T_b = 362$ K), we get

$$T_f = \frac{(0.32 \text{ kg})(4186 \text{ J·kg}^{-1}\text{·K}^{-1})(295 \text{ K}) + (0.26 \text{ kg})(900 \text{ J·kg}^{-1}\text{·K}^{-1})(362 \text{ K})}{(0.32 \text{ kg})(4186 \text{ J·kg}^{-1}\text{·K}^{-1}) + (0.26 \text{ kg})(900 \text{ J·kg}^{-1}\text{·K}^{-1})} = 305 \text{ K}$$
$$(T7.2$$

(See problem T1S.7 for another way to compute the equilibrium temperature.) Since $T_f = 305$ K = 42°C is significantly different from the water's initial temperature 22°C, we need equation T7.19 to compute the water's entropy change:

$$\Delta S_w = MC \ln\left(\frac{T_f}{T_w}\right) = (0.32 \text{ kg})(4186 \text{ J·kg}^{-1}\text{·K}^{-1}) \ln\left(\frac{305 \text{ K}}{295 \text{ K}}\right) = +45 \text{ J/K.} \qquad (T7.23)$$

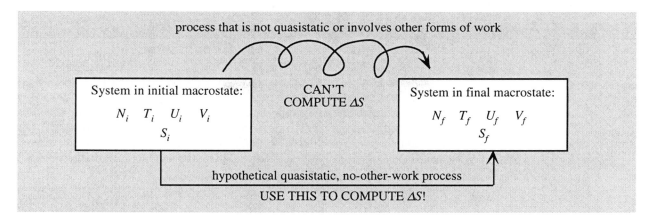

Figure T7.2: How to do an end-run around a process violating the restrictions on equation T7.6: invent a process that goes from the same initial to the same final macrostate and use *that* process to compute ΔS. Since ΔS only depends on the initial and final macrostates, it doesn't matter what process you use to compute it.

Exercise T7X.6: Did we have to use absolute temperatures in equation T7.22 (that is, would using temperatures in °C make any difference in the final result)? Did we have to use absolute temperatures in equation T7.23?

Exercise T7X.7 Show that the entropy change of the block is –40 J/K. (This means that the net entropy change for the block/water system is +5 J/K: an *increase* is to be expected in a spontaneous heat transfer process.)

T7.7 NON-QUASISTATIC PROCESSES

As I've said before, our basic equations T7.6 and T7.7 don't apply to processes where volume changes are not quasistatic or if work other than that resulting from quasistatic volume changes are done during the process. In many cases of such processes, however, we can actually use these equations, using a clever argument to do a kind of "end run" around the process.

Our clever argument hinges on the fact that the entropy of any system is defined in terms of the current *macrostate* of that system. Therefore any change in entropy experienced by the system as it goes from one macrostate to another *will depend on its initial and final macrostates alone*, and not at all on the process by which we got from one to the other!

An object's entropy depends on its macrostate alone

Consider a system that goes from one macrostate to another via a process that is *not* quasistatic, or involves work *not* due to a quasistatic expansion or compression, or both. Say that we'd like to compute the entropy change of the system in this process. If we can imagine *any* hypothetical process satisfying the restrictions on equation T7.6 that takes us from the same initial macrostate to the same final macrostate, we can calculate the entropy change using this hypothetical process (which we will call a **replacement process**) instead of the actual process: we *must* get the same answer that we would have gotten if we could have done the calculation for the actual process (see Figure T7.2).

Using a *replacement process* to calculate an entropy change

Problem: Imagine that I have a basin containing 250 kg of water at 22°C, and I pour a pitcher containing 5 kg of water at 22°C into it from a height of 1.5 m. What is the entropy change of the system of both containers of water?

EXAMPLE T7.5

Solution When the 5 kg of water falls into the basin, its gravitational potential energy is converted to kinetic energy and then to thermal energy in the basin. This thermal energy gain is *not* due to a quasistatic volume change, nor is it heat entering the system (there are no temperature differences involved). Therefore, equation T7.6 technically does not apply.

EXAMPLE T7.5 (cont.)   The initial macrostate of our system is 255 kg of water (250 kg in the basin and 5 kg in the pitcher) at 22°C and thermal energy U_i. Our final state is the same of water at a slightly higher temperature and a final thermal energy of

$$U_f = U_i + mgh = U_i + (5 \text{ kg})(9.8 \text{ m/s}^2)(1.5 \text{ m}) = U_i + 74 \text{ J} \qquad \text{(T7.24)}$$

Imagine gently lowering the pitcher to the level of the basin and mixing the water in slowly and gently (avoiding adding *any* work as much as possible) and then adding 74 J of heat energy to the water. This replacement process will bring the water to the *same* final state without involving any non-quasistatic work.

Since it would take $(255 \text{ kg})(4186 \text{ J·kg}^{-1}\text{·K}^{-1})(1 \text{ K}) > 10^6$ J to raise the temperature of the water by even 1 K, our 74 J will not change the water's temperature significantly. Therefore, we can use equation T7.6 to compute the water's entropy change during the replacement process:

$$\Delta S = +\frac{\Delta Q}{T} = \frac{74 \text{ J}}{(273 + 22) \text{ K}} = +0.25 \text{ J/K} \qquad \text{(T7.25)}$$

The water's entropy change during the actual process *must* be the same.

EXAMPLE T7.6   **Problem:** Imagine that we confine 1.0×10^{23} molecules of a gas at 22°C to *half* of an insulated chamber whose total volume is 0.50 m^3 (the other half contains a vacuum). We then rupture the wall between the halves. What is the entropy change associated with the gas spontaneously filling the vacuum?

Solution Since the process is spontaneous and irreversible, we know that the entropy of the gas must have increased. However, the volume change here is not even remotely quasistatic, so equation T7.6 does not directly apply.

Note that the gas expands freely into the vacuum, without doing any work (as it would if it were to push against a piston). Moreover, the expansion occurs suddenly and the chamber is insulated, so there will be neither time nor means for heat to flow into or out of the system. The total thermal energy of the gas is thus unchanged. The initial macrostate of the gas is thus specified by the quoted value of N, some (unknown) thermal energy U, and some initial volume V_i. Its final macrostate has the same N, the same U, and a final volume $V_f = 2V_i$.

One replacement process that takes us to the same final macrostate is the following. Imagine that we use a piston to gradually allow the gas to expand to its final volume. As the gas expands against the piston, though, it will do *work*, decreasing the thermal energy of the system. We want the final energy of the gas to be the same as it was originally, so we need to add heat energy to replace the work energy lost. We can do this in a particularly easy way by putting the gas in thermal contact with a reservoir at 22°C. Since the thermal energy of an ideal gas depends on T and N but not V, keeping the gas' temperature fixed will automatically add whatever heat is needed to keep its internal energy fixed as well.

The heat added in this replacement process will be equal to the work that the gas does as it expands isothermally. According to equation T3.10b, then

$$\Delta Q = -W = +Nk_BT\ln\!\left(V_f / V_i\right) \qquad \text{(T7.26)}$$

Since the temperature of the gas is a constant 22°C = 295 K in our replacement process, we can use equation T7.6 to calculate the gas' change in entropy:

$$\Delta S = \frac{\Delta Q}{T} = +\frac{Nk_BT}{T}\ln\!\left(\frac{V_f}{V_i}\right) = Nk_B\ln(2) = +0.96 \text{ J/K} \qquad \text{(T7.27)}$$

This has to be the entropy change in the actual process as well.

Exercise T7X.8: Check the numerical result in equation T7.27.

Exercise T7X.9: You stir a 250-g cup of coffee at 85°C, doing about 0.03 J of work on it as you stir. What is the entropy change of the coffee?

I. ENTROPY AND TEMPERATURE   **SUMMARY**
   A. The definition of temperature is based on three principles:
      1. S_{TOT} of a system of interacting objects is maximum in equilibrium
      2. the total energy of an isolated system is conserved
      3. temperatures are equal in equilibrium by the zeroth law
   B. These principles suffice to establish that $dS/dU = f(T)$
      1. then we choose $f(T)$ to be $1/T$ to coincide with gas temperature
      2. we'd probably choose this anyway: this is simplest $f(T)$ that
         a) makes sure that hot objects have numerically high T's and
         b) makes sure thermal energy per molecule is approximately $\propto T$
   C. But $dS/dU \equiv 1/T$ *assumes* that N, V, and other variables are fixed

II. THE CLAIM THAT $\Delta S = 0$ FOR ADIABATIC VOLUME CHANGES
   A. Imagine an ideal gas held in an insulated cylinder by a frictionless piston
   B. If we set the piston oscillating, I claim it will oscillate forever. Why?
      1. for an adiabatic process, $PV^\gamma = $ constant
      2. so the pressure is the same at a given piston position every cycle
      3. so the gas is like a spring whose force also depends on position
   C. An indefinite oscillation is reversible, so the gas' entropy is fixed
      1. note that the gas is insulated from its surroundings
      2. changes in piston's *mechanical energy* doesn't change its entropy
      3. therefore, there are no entropy changes *outside* the gas either
   D. Conclusion: $\Delta S = 0$ for any quasistatic, adiabatic volume change

III. CALCULATING ENTROPY CHANGES
   A. About the work done during a quasistatic volume change
      1. If N and V are fixed, then $dS = dU/T$ (from definition of T)
      2. In an adiabatic volume change, $dU = $ work, but $dS = 0$
      3. This suggests that the work energy flowing into a gas due to a quasistatic volume change doesn't actually change its entropy
      4. Suggests that $dS = (dU - dW_{qs})/T$ for general ideal gas process
      5. If no *other* kinds of work are done, $dU = dQ$ and $dS = dQ/T$ (T7.6)
   B. The latter formula applies to solids and liquids, too!
      1. N and V are essentially fixed for solids and liquids, so $dS = dU/T$
      2. if no other kinds of work are involved, then $dU = dQ$, $dS = dQ/T$
   C. Constant temperature processes:
      1. $dS = dQ/T$ assumes that dQ "sufficiently small" so that $T \approx$ const.
      2. this happens in real life in three typical cases:
         a) the object is a *reservoir* (so large that $dT \approx 0$ for given dQ)
         b) the object is in thermal contact with a reservoir
         c) the object is changing phase ($T = $ constant until complete)
      3. In such cases, dQ *is* "sufficiently small": use equation T7.6 directly
   D. Heat exchanges where temperature changes significantly:
      1. if T not constant, we have to integrate: $\Delta S = \int dQ/T$
      2. when no work is involved and the specific heat* is $\approx$ constant
         a) no work implies $dQ = dU = mcdT$, where $c = $ specific heat*
         b) so $\Delta S = \int mc\, dT/T = mc \int dT/T = mc \ln (T_f/T_i)$ (T7.19)
   E. Non-quasistatic processes (equation T7.6 doesn't apply directly!):
      1. S depends only on U, V, N, ...; that is, on *macrostate*, not history!
      2. ΔS is the same for *any* process going between the same macrostates
      3. so find ΔS for a *replacement process* going between the same states

GLOSSARY

quasistatic (volume change of a gas): a volume change that occurs slowly enough so that the gas is almost in equilibrium at all times and has a well-defined pressure whose value is the same throughout the gas at all times.

adiabatic: a process that occurs without heat flowing into or out of the system in question.

mechanical energy: the kinetic energy of an object's macroscopic and potential energy of its external interactions, as contrasted with the *internal energy* it stores at the microscopic level. Increasing an object's mechanical energy does not change its entropy.

reservoir: an object so massive that it can absorb or provide substantial amounts of heat without a discernable change in its temperature.

phase change: the transformation of a substance from one *phase* to another. Substance phases most commonly seen are *solid*, *liquid*, and *gas*.

latent heat* L: a substance's change in thermal energy per unit mass during a phase change: $\Delta U = \pm mL$. ΔU is positive if the substance goes from solid to liquid, liquid to gas, or solid to gas and negative if the transformation goes the other direction.

replacement process: a hypothetical quasistatic process that we use to compute the entropy change of a system during a process involving non-quasistatic work. Since a system's entropy depends only on its macrostate, if we can think of a quasistatic replacement process that takes the system between the same initial and final macrostates as the non-quasistatic process, the entropy change for both processes should be the same.

TWO-MINUTE PROBLEMS

T7T.1 Consider each of the following processes. In which does the total entropy of the interacting objects increase, and in which does it remain (essentially) the same? Answer C if the entropy remains constant, D if there is a change.
(a) a coin drops to the ground
(b) an ideal gas is compressed in an insulated container
(c) a pendulum swings back and forth in a vacuum
(d) the earth orbits the sun
(e) a stream runs down a mountainside

T7T.2 A ball is thrown vertically into the air in a vacuum. What happens to the entropy of the ball during the rising part of its trajectory? What happens to its entropy during the falling part of its trajectory? The ball's entropy
A. doesn't change at all
B. increases as it rises, decreases as it falls
C. decreases as it rises, increases as it falls
D. increases throughout the process
E. decreases throughout the process

T7T.3 Consider each of the following processes. Is the process consistent (C) or inconsistent (D) with the limitations of equations T7.6 and T7.7 (that N be fixed and that the process involve no work *other* than the work produced by a quasistatic volume change)?
(a) a glass of milk is spilled on the floor
(b) a closed vial of gas is heated with a flame
(c) a pan of water boils on the stove
(d) sunlight is absorbed by a black hat
(e) water evaporates from your skin

T7T.4 100 kg of water in a tub at 30°C absorbs 100 J from an electric heater. By about how much does its entropy increase in this process?
A. 0.033 J/K   D. 33 J/K
B. 0.33 J/K   E. 1400 J/K
C. 3.3 J/K   F. other

T7T.5 100 g of water is slowly heated from 5°C to 25°C. What is its change in entropy? (*c* for water = 4186 J/K)
A. 582 J/K   D. 6.95 J/K
B. 29.2 K   E. zero
C. 139 J/K   F. other (specify)

T7T.6 A hot object is brought into contact with a cold object. The hot object's entropy decreases by 1.5 J/K as it comes into equilibrium with the cold object. How much does the cold object's entropy increase?
A. more than 1.5 J/K   C. less than 1.5 J/K
B. exactly 1.5 J/K   D. zero

T7T.7 A sample of helium gas is held at a constant temperature while it is slowly compressed in a cylinder until its volume has decreased by a factor of four. During this process, the gas' entropy
A. increases   C. remains the same
B. decreases   D. other (specify)

T7T.8 A stone of mass m is dropped from a height h into a bucket of water. What would be a suitable replacement process for this non-quasistatic process?
A. Lower the stone gently into the water.
B. Lower the stone gently into water and then heat the water until it has mgh more energy than it had before.
C. Lower the stone gently into the water and then stir the water gently to give it mgh more energy.
D. Raise the water up to height h, gently put the stone in it, and then lower both back to the ground.
E. Either B or C   F. none of the above

T7T.9 A sample of helium gas in an insulated cylinder is suddenly and violently compressed to half its volume. The absolute temperature of the gas after the compression is observed to have doubled. (Note: in a *quasistatic* adiabatic compression of a monatomic gas, $TV^{2/3} = $ const). A suitable replacement process for this process would be
A. a slow isothermal compression from V_{init} to $\frac{1}{2}V_{init}$.
B. a slow adiabatic compression from V_{init} to $\frac{1}{2}V_{init}$.
C. heating the gas to double its temperature followed by a slow adiabatic compression from V_{init} to $\frac{1}{2}V_{init}$.
D. a slow adiabatic compression from V_{init} to $\frac{1}{2}V_{init}$ followed by heating to bring the temperature to $2T_{init}$.
E. a slow adiabatic compression from V_{init} to $\frac{1}{2}V_{init}$ followed by cooling to bring the temperature to $2T_{init}$.
F. none of the above.

HOMEWORK PROBLEMS

BASIC SKILLS

T7B.1 Explain in your own words how you know that the entropy of a system consisting of a mass oscillating up and down at the end of a spring is nearly constant, while the entropy of a system consisting of an object sliding to a rest on a rough surface clearly increases.

T7B.2 Explain in your own words how you know that the entropy of a swinging pendulum is nearly constant, while the entropy of two balls of putty that collide and stick together clearly increases.

T7B.3 Imagine that we add 12 J of heat to 10 kg of water at 15°C. Argue that this amount of energy will not change the temperature of the water very much. Compute the entropy change of the water.

T7B.4 Imagine that we put a small ice cube in a bucket containing 5 kg of water at 22°C. As the ice melts, it absorbs 35 J of energy from the water. By about how much has the entropy of the water decreased?

T7B.5 Imagine that the temperature of a 220-g block of aluminum sitting in the sun increases from 18°C to 26°C. By about how much has its entropy increased?

T7B.6 Imagine that a 330-g cup of water is heated in a hot pot from 12°C to 92°C. By about how much has its entropy increased?

T7B.7 125 g of water at 100°C in a pan on a stove is converted into steam. What is the entropy change of the water that is now steam? (The latent heat* of boiling water is 2256 kJ/kg.)

T7B.8 A puddle containing 0.80 kg of water at 0°C freezes on a cold night, becoming ice at 0°C. What is the entropy change of the water that is now ice? (The latent heat* of freezing water is 333 J/K.)

SYNTHETIC

T7S.1 Imagine that you put 180 g of ice in a cold drink that is already at 0°C. After a few minutes about half the ice is melted and the drink is still at 0°C. What is the entropy change of the ice that melted?

T7S.2 Imagine that you use an electrical wire to add 28 J of energy to a basin containing 65 kg of water at 18°C. What is the approximate entropy change of the water? [HINT: argue that the water is essentially a "reservoir".]

T7S.3 Imagine that you put a 1.0 kg block of aluminum whose initial temperature is 80°C into the ocean at a temperature of 5°C. The ocean and block come into thermal equilibrium. What was the entropy change of the block in this process? The ocean?

T7S.4 Imagine that 22 g of helium gas in a cylinder expands quasistatically while in contact with a reservoir at a temperature of 25°C and does 85 J of work in the process. What is the entropy change of the gas? The reservoir?

T7S.5 Imagine that you quasistatically compress 1.0 mole of oxygen gas by a factor of two while it is in contact with a reservoir at 18°C, and you do 120 J of work in the process. What is the entropy change of the gas?

T7S.6 Imagine that you compress 1.0 mole of helium gas from a volume of 0.120 m³ to a volume of 0.040 m³ while the gas is held at a constant temperature of 278 K. By how much does the entropy of the gas change?

T7S.7 Imagine that you allow 0.40 mole of nitrogen to expand from a volume of 0.005 m³ to a volume of 0.015 m³ while the gas is held at a constant temperature of 304 K. By how much does the entropy of the gas change?

T7S.8 Imagine that you put a block of copper with a mass of 320 g and an initial temperature of 85°C into an insulated cup of water containing 420 g of water at 0°C. Compute the entropy changes of both the water and the copper.

T7S.9 Imagine that you put a block of copper with a mass of 320 g and an initial temperature of −35°C into an insulated cup of water containing 420 g of water at 22°C. After everything has come to equilibrium, by how much has the entropy of both the water and the copper changed? Explain why the coat of ice that initially forms around the copper is irrelevant.

T7S.10 Imagine that you have a gas confined in an insulated cylinder by a piston. Imagine that you suddenly push hard on the piston, compressing the gas suddenly to half its volume. Then you slowly let the gas expand back to its original volume. Will the gas have the same temperature as when you started? Will its entropy be the same as before? Would your answers be different if you had slowly compressed the gas and then slowly expanded it? Explain your responses carefully. (*Hint:* when you press the piston in suddenly do you think that you exert more, less, or the same force on it that you would if you were to compress the gas slowly?)

T7S.11 A 3-g bullet flying at 420 m/s hits a 2.2-kg aluminum block and imbeds itself in the aluminum. After everything has come to equilibrium, how much has the entropy of the aluminum block increased? Be sure to describe the replacement process that you use to actually calculate the entropy change.

RICH-CONTEXT

T7R.1 A high diver dives from the top of a 35 m tower into a pool of water. After the splashing settles down (but before any heat has a chance to be transferred from the water to the diver or vice versa) what is the entropy change of the water in the pool? Explain why one can't calculate the entropy change for this process directly, and describe the replacement process that you used to calculate the entropy change. Also describe any approximations or estimations that you have to make.

T7R.2 A small amount of gasoline explodes in an automobile engine cylinder, releasing 350 J of energy into about 0.060 moles of air at a temperature of 480 K compressed within a volume of 120 cm³. Subsequently the hotter air pushes the piston down, expanding the gas roughly adiabatically to 480 cm³. What is the approximate entropy change of the air in this process (starting just before the gas exploded)? Explain why one can't calculate the entropy change for this process directly, and describe the replacement process that you used to calculate the entropy change.

ADVANCED

T7A.1 This problem explores a mathematical way to show that the entropy change during an adiabatic volume change is zero. In the problem T6S.7, I claim that the multiplicity of an ideal monatomic gas is given by

$$g(N,U,V) = CV^N U^{3N/2} \qquad (\text{T7.28})$$

where C is a constant that does not depend on V or U. In parts (a) and (b) of this problem, I will argue that based on what we know about the ideal gas, this pretty much *has* to be the correct multiplicity function. In part (c), I will argue that this multiplicity function implies that an adiabatic volume change will not change the gas' entropy.

(a) The entropy for this multiplicity function is

$$S \equiv k_B \ln g = k_B[\ln C + N\ln V + \tfrac{3}{2}N\ln U] \qquad (\text{T7.29})$$

Show from the definition of temperature that this entropy function implies that the gas' thermal energy at a temperature T will be $U = \tfrac{3}{2}Nk_BT$. This result is consistent with what we measure for actual monatomic gases (like helium). No other function of U other than $U^{3N/2}$ in the multiplicity will give this experimentally-supported result. Thus the multiplicity had better include a factor of $U^{3N/2}$.

(b) We also argued in section T6.4, that if the volume available to a gas increases by a factor of 2, the number of microstates available to the gas must increase by a factor of 2^N. *Explain why.* Similarly, if the volume increases by a factor of 3, the multiplicity will increase by a factor of 3^N and so on. Therefore, the multiplicity should have a factor of V^N in it, and this should be the only way that the multiplicity depends on V.

(c) Show using the entropy function in equation T7.29 that the entropy will be constant during a volume change (with N fixed) *if and only if*

$$UV^{2/3} = \text{constant} \qquad (\text{T7.30})$$

(Alternatively, you can argue that g given by equation T7.28 will be fixed *if and only if* this is true: if g is fixed then S is fixed.) For an ideal gas, though, $T \propto U$, so this equation is equivalent to saying that

$$TV^{2/3} = \text{constant} \qquad (\text{T7.31})$$

Use the ideal gas law to show that this means that

$$PV^{5/3} = \text{constant} \qquad (\text{T7.32})$$

This is the law of adiabatic expansion given in chapter T3 ($\gamma = 5/3$ for monatomic gases). Therefore, we have shown that if T7.32 holds, $\Delta S = 0$, and for ΔS to be zero (with N fixed) equation T7.32 must hold. Therefore $\Delta S = 0$ for our gas *if and only if* its volume change is adiabatic.

ANSWERS TO EXERCISES

T7X.1 At a certain place in the argument, I noted that in the adiabatic case, the insulation thermally isolates the gas from its surroundings, meaning that the gas cannot affect the entropy of its surroundings or vice versa. This was part of my argument that the total system's entropy change during the oscillation being zero implies that the *gas'* entropy change during the oscillation is also zero. If the gas is in thermal contact with its surroundings, however, we cannot make this argument: heat flowing into or out of the gas can clearly change the entropy of its surroundings. So in this case, even if the system as a whole undergoes zero change in entropy, we cannot be sure that the gas does not undergo a change in entropy that is canceled by an opposite change in its exterior.

T7X.2 Consistent: *b,d,e.* Inconsistent: *a, c.*

T7X.3 –133 J/K. Yes, the total entropy of the system increases by about 17 J/K. (We would expect an entropy increase, since the process is irreversible.)

T7X.4 Equation T7.6 is based on the definition of temperature (equation T7.1). In the chapter, we set up this definition to coincide with the absolute temperature scale defined by the ideal gas thermoscope. So T in any equation based on equation T7.1 has to be *absolute* temperature.

T7X.5 Answer is given. Note that:

$$\int_{T_i}^{T_f} \frac{dT}{T} = \ln T_f - \ln T_i = \ln\!\left(\frac{T_f}{T_i}\right) \qquad (\text{T7.33})$$

T7X.6 Since equation T7.22 is based on a rearrangement of equation T7.20, where only temperature *differences* appear, it does not matter whether we use temperatures in °C or kelvins (1 K = 1°C of temperature difference). In equation T7.23, on the other hand, we must use absolute temperatures. (I generally try to use absolute temperatures exclusively to reduce my chances of error.)

T7X.7 The block's change in entropy is

$$\Delta S_b = mc\ln\!\left(\frac{T_f}{T_b}\right)$$

$$= (0.26\ \text{kg})(900\ \text{J·kg}^{-1}\text{·K}^{-1})\ln\!\left(\frac{305\ \text{K}}{362\ \text{K}}\right) = -40\ \text{J/K} \qquad (\text{T7.34})$$

T7X.8 (The answer is given.)

T7X.9 The entropy change is $\approx +8.4\times10^{-5}$ J/K = 84 μJ/K.

HEAT ENGINES

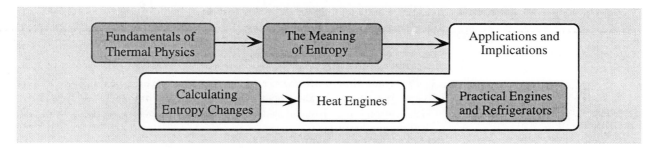

T8.1 OVERVIEW

A **heat engine** is a device that is capable of continuously converting heat energy into mechanical energy. The Industrial Revolution began in the mid-1700's when human beings first figured out how this might be done. Heat engines still dominate our industrial economy: internal combustion engines, steam turbines, and even jet engines and guns are basically heat engines. Heat engines have been so essential to all aspects of technology (and thus culture) that it might be argued that heat engines have done more to shape history than any single invention since agriculture. Since virtually all heat engines in use today consume fossil fuels and thus contribute to the greenhouse effect, heat engines may well continue to shape our future (perhaps even long after the last drop of fossil fuel is gone).

In the last chapter, we learned how to compute entropy changes in various kinds of thermodynamic processes. In this chapter, we will apply our skills to learning about heat engines in abstract: why it is possible to construct a heat engine at all, how heat engines work, and what limits the second law places on their operation and efficiency. In chapter T9, we will draw on ideas developed in chapter T3 to discuss how to actually construct heat engines and refrigerators.

Here is an overview of the sections in this chapter.

T8.2 *PERFECT ENGINES ARE IMPOSSIBLE* discusses why a perfect heat engine (one that converts heat energy *entirely* into mechanical energy) is a violation of the second law of thermodynamics.

T8.3 *REAL HEAT ENGINES* argues that even so, we *can* construct *imperfect* heat engines (that convert *some* heat into mechanical energy).

T8.4 *THE EFFICIENCY OF A HEAT ENGINE* explores the stringent limitations that the second law of thermodynamics puts on the efficiency of such engines (no matter how cleverly they are designed).

T8.5 *CONSEQUENCES* examines some consequences of this limitation.

T8.6 *REFRIGERATORS* discusses how it is possible for a refrigerator to move heat energy from cold to hot against its natural direction of flow.

One of the beauties of the second law is that we can say some definite and important things about heat engines and refrigerators without ever delving into the details of how they actually work (that will be our focus in the next chapter)! The things that we uncover in this chapter therefore apply to *any* kind of heat engine or refrigerator, even ones not yet invented!

T8.2   PERFECT ENGINES ARE IMPOSSIBLE

Perpetual motion machines

Since the dawn of the Industrial Revolution (and perhaps before) inventors have been trying to build a **perpetual motion machine**: a machine that would run endlessly without using any fuel. (Obviously, if such a machine could be constructed, it would make its inventor very rich.) Many attempts to achieve perpetual motion have been made in the past several centuries, some quite ingenious. But all such attempts have failed, and indeed *must* fail. Why?

All designs for perpetual motion machines fall into two broad categories:

1. *Perpetual Motion Machines of the first type.* These designs seek to create the energy required for their operation out of nothing, thus violating the law of Conservation of Energy (the first law of thermodynamics).

2. *Perpetual Motion Machines of the second type.* These designs extract the energy required for their operation from sources in a manner that requires the entropy of an isolated system to decrease (thus violating the second law of thermodynamics.)

You may have seen proposed designs for perpetual motion machines of the first type. An example would be a self-powered electric car that would use its motion through the air to turn a windmill that would recharge the car's batteries, thus allowing it (in principle) to run forever. But since a moving car will always dissipate energy (due to various forms of friction), and since no source for that energy is apparent in this design, it must fail.

Perpetual motion machines of the second type are usually a bit less obviously flawed. For example, the earth's oceans contain an enormous amount of internal thermal energy. Why not construct a ship engine that soaks up some of this internal energy (making the ocean a bit cooler) and converts it to mechanical energy to drive the ship forward? (Such an engine would obviously save shippers a tremendous amount of money!) Conservation of energy is not violated here, so it seems more plausible that such an engine might work. Yet such an engine violates the *second* law of thermodynamics. Why?

Transfers of mechanical energy do not necessarily change an object's entropy

The key to answering this question is to understand that *transferring purely mechanical energy to or from an object does not (in general) change its entropy.* For example, the engine powering a crane that lifts a massive girder transfers quite a bit of mechanical energy to the rising girder, but this energy does not increase the girder's entropy: it has the same number of microstates available to it when it is lifted into the air that it had when it was sitting on the ground. An engine that accelerates an automobile increases that automobile's kinetic energy, but this does not raise the entropy of the automobile, which has exactly as many microstates when it is moving as it has when it is at rest (other things being the same). We discussed in the last chapter how the adiabatic quasistatic expansion or compression of an ideal gas can transfer work energy from or to the gas without changing its entropy. *Transfers of mechanical energy do not cause entropy changes.*

I also claim that any engine that is capable of continuously converting heat to mechanical energy cannot increase its own entropy indefinitely. Virtually all heat engines operate in cycles, where a working substance (typically a gas) is manipulated so that it absorbs energy during part of the cycle, expands against a piston to produce mechanical energy later in the cycle, and then is returned to its original state in the last part of the cycle, ready to convert the next batch of heat energy. It is crucial that the working substance be returned to its original state if the engine is going to operate in the same manner continuously. But if the macrostate of the working substance is the same at the end of a cycle as it was during the beginning, its net entropy change is also zero. Therefore, the net entropy change during a cycle of a continuously operating engine must be zero.

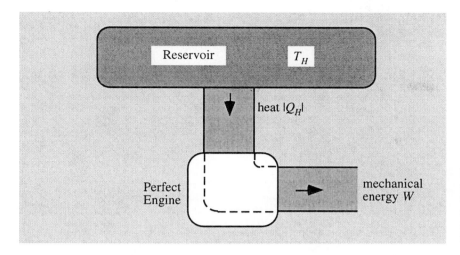

Figure T8.1: A schematic diagram of a perfect heat engine, which extracts a certain amount of heat energy from a reservoir and converts it entirely into useful mechanical energy.

The handful of heat engines that do not operate in cycles operate in a steady state, where the working substance has the same macrostate all the time, once steady operation has been established. The entropy of such a steady-state engine also clearly does not change in time.

Figure T8.1 shows an abstract diagram of a "perfect" heat engine (such as our hypothetical ocean-powered engine). In a given interval of time (say, one cycle of the engine), such an engine would extract heat energy $|Q_H|$ from a thermal reservoir at temperature T_H (such as the ocean) and turn that energy entirely into mechanical energy, as shown. (The schematic diagram illustrates the flow of energy almost as if it were a river of water. This is a useful mental image: since energy is conserved, it does behave much like a flowing indestructible substance.)

This diagram helps us to understand why such an engine violates the second law of thermodynamics. The net entropy change of the engine during a complete cycle is zero. The mechanical energy produced by the engine also does not produce any entropy changes. But the process of extracting heat energy from the reservoir *decreases* its entropy. Since the temperature of the reservoir remains constant, the entropy decrease is in fact easy to calculate:

Why a perfect engine violates the second law

$$\Delta S_H = \frac{Q_H}{T_H} = -\frac{|Q_H|}{T_H} \tag{T8.1}$$

where ΔS_H is the entropy change of the reservoir, T_H is its (constant) temperature, and Q_H is the heat flowing out of the reservoir (which is negative because heat is flowing *out* of the reservoir).

Exercise T8X.1: The second law of thermodynamics states that *the entropy of an isolated system of interacting objects must increase.* So exactly how does our hypothetical perfect engine violate this law?

T8.3 REAL HEAT ENGINES

If extracting heat from a reservoir causes a forbidden decrease in entropy, then how is it possible to construct a heat engine at all? The trick is essentially as follows. We saw in the last chapter that when heat flows from a hot object to a cold object, the entropy of the combined system of the two objects increases by some finite amount: the entropy of the hot object decreases as energy flows out, but the entropy of the cold object increases by a greater amount as the same energy flows in. A heat engine takes this basic flow of energy from hot to cold

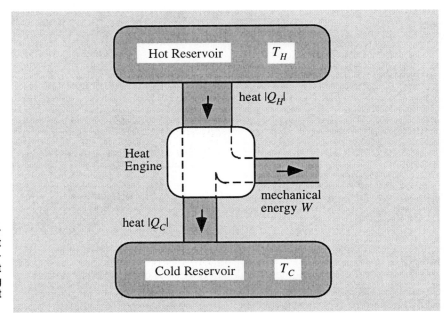

Figure T8.2: A schematic diagram of a real heat engine. A heat engine essentially taps the spontaneous flow of heat from a hot object to a cold object, diverting some of the flow and converting it into mechanical energy.

and diverts some of it, converting the diverted portion into mechanical energy. As long as we don't divert too much, there will still be enough energy flowing into the cold object to keep its entropy increase greater than the entropy loss experienced by the hot object.

A visual representation of what a heat engine does

Figure T8.2 is a schematic illustration of what a heat engine does. The presence of a *temperature difference* driving the energy flow is essential. In anthropomorphic language, heat wants so badly to flow from hot to cold, that one can make it do a little bit of work in the process. What makes heat energy "want" to flow is the increase of entropy that results from that flow (there are all these wonderful microstates in the cold reservoir that could be sampled with just a bit more energy). We can divert some of this flowing energy to mechanical energy as long as we don't divert so much that the original "purpose" of the energy flow is removed.

An example heat engine: a steam turbine

A good example of a heat engine is a steam turbine in an electrical power plant (see Figure T8.3). Heat from some source (nuclear power or burning fossil fuels) converts water in a boiler to steam. (Since heat is constantly supplied to the boiler at a constant temperature by whatever the heat source is, the heat source acts as if it were a constant-temperature hot reservoir.) Water expands greatly when it boils, so this means that the steam is under high pressure. This high-pressure steam is allowed to expand by blowing against the blades of a turbine, which converts some of the internal energy of the steam into mechanical energy (which is ultimately turned into electrical energy by a generator). The spent steam is sent to a condenser, which cools the steam and condenses it back into liquid water by putting it in contact with a cold reservoir. In the old days, a nearby river was used as the cold reservoir, but this often increased the temperature of the water enough to be harmful to life downstream. Most power plants built recently employ cooling towers, which essentially use the atmosphere as the cold reservoir. In either case, the recondensed water is then returned to the boiler in essentially its original state, completing the cycle.

Any temperature difference can (in principle) be used to drive a heat engine. During the energy crisis of the 1970s, some research was done on the possibility of generating electrical energy from the temperature difference between the ocean surface and ocean floor (which can be as large as 30 K). The plan for such a power plant would be much like that shown in Figure T8.3, except that we would have to use a working substance that boils at a lower temperature than water

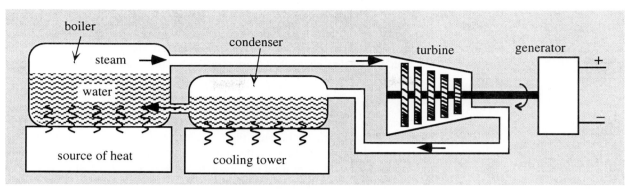

boiler
condenser
turbine
generator
steam
water
source of heat
cooling tower
+
−

FigureT8.3: A schematic diagram of a steam turbine in an electrical power plant. Something (burning fossil fuels, nuclear power) acts as a *hot reservoir*, whose heat boils water, converting it to steam which expands against the blades of the turbine, turning the generator and creating electrical energy. The spent steam is sent to a *condenser* where it gives up heat to the atmosphere (the cold reservoir) and condenses back into water, completing the cycle.

(ammonia is a possibility). Such a plant would generate electrical power without using any fuel (ultimately, the temperature difference in the ocean is maintained by energy from the Sun)! There are, however, some thorny practical and environmental issues associated with doing this on a large scale. (A good discussion of this proposal can be found in Penney and Bharathan, "Power from the Sea", *Scientific American*, January 1987.)

Exercise T8X.2: I claim that in a heat engine such as the one schematically illustrated in Figure T8.2, the mechanical energy W produced in a cycle of the engine is related to the heat extracted from the hot reservoir $|Q_H|$ and heat absorbed by the cold reservoir $|Q_C|$ by the equation:

$$W = |Q_H| - |Q_C| \qquad \text{(T8.2)}$$

Explain why.

T8.4 THE EFFICIENCY OF A HEAT ENGINE

The purpose of any heat engine is to convert into mechanical energy as much of the heat $|Q_H|$ extracted from the hot reservoir (which is typically energy provided by burning precious fuel) as possible. We can quantify how well an engine does this in terms of its **efficiency** e, which is defined to be the ratio of the mechanical energy W produced to the heat energy $|Q_H|$ extracted:

$$e \equiv \frac{W}{|Q_H|} \qquad \text{(T8.3)}$$ **Definition of efficiency**

The second equality follows from equation T8.2.

Exercise T8X.3: Combine equations T8.2 and T8.3 to show that the efficiency can be written

$$e = 1 - \frac{|Q_C|}{|Q_H|} \qquad \text{(T8.4)}$$

Note that a perfect engine of the type described in section T8.2 would have an efficiency of $e = 1$: such an engine converts all of $|Q_H|$ into W and gives up no heat $|Q_C|$ to a cold reservoir. As we've seen, such a perfect engine violates the second law of thermodynamics, and so is impossible. The efficiency of any real engine will therefore be less than 1.

The limit on efficiency imposed by the second law

The question is, how much less? We can in fact calculate this fairly easily. Imagine that one cycle of our heat engine absorbs heat $|Q_H|$ from a hot reservoir at a fixed temperature T_H, produces a certain amount of mechanical energy W, and then gives up heat $|Q_C|$ to a cold reservoir at a fixed temperature T_C. Since both the hot and cold reservoirs are considered to have fixed temperatures, we can calculate the entropy changes of these reservoirs using equation T7.8:

$$\Delta S_H \ = \ -\frac{|Q_H|}{T_H}, \qquad \Delta S_C \ = \ +\frac{|Q_C|}{T_C} \tag{T8.5}$$

Note that ΔS_H is *negative* because the hot reservoir loses entropy as it gives up heat. The cold reservoir, on the other hand, *gains* entropy as it absorbs heat. In the formula above, I have used the absolute value of the heat lost or gained to make the signs of ΔS_H and ΔS_C explicit.

Exercise T8X.4: The entropy change of the engine itself during a cycle is zero, and transferring mechanical energy does not intrinsically cause any entropy changes. Argue that requiring that the entropy of the isolated interacting system consisting of the engine and the reservoirs must increase (or at least remain the same) implies that:

$$\frac{|Q_C|}{T_C} - \frac{|Q_H|}{T_H} \ \geq \ 0 \tag{T8.6}$$

Exercise T8X.5: Show that this can be rearranged to read:

$$\frac{|Q_C|}{|Q_H|} \ \geq \ \frac{T_C}{T_H} \tag{T8.7}$$

Exercise T8X.6: Combine equations T8.4 and T8.7 to show that:

$$e \ \leq \ 1 - \frac{T_C}{T_H} \ = \ \frac{T_H - T_C}{T_H} \tag{T8.8}$$

This equation says that any heat engine operating between two reservoirs at temperatures T_H and T_C respectively has a maximum possible efficiency of:

Maximum theoretical efficiency of a heat engine

$$e_{\text{max}} \ = \ \frac{T_H - T_C}{T_H} \tag{T8.9}$$

no matter how cleverly it is designed. Any engine that seeks to do better than this will give up too little heat to the cold reservoir to make its entropy gain exceed the hot reservoir's entropy loss, and thus will violate the second law.

Note that this maximum efficiency is essentially equal to the fraction that the temperature *difference* between the reservoirs is of the temperature of the high-temperature reservoir. As the temperature difference between the reservoirs increases relative to T_H, the maximum possible efficiency goes up. As the temperature of the cold reservoir approaches absolute zero, the efficiency of the engine approaches the perfect value of 1.

The derivation of equation T8.8 described here is one of those derivations that every educated person should know and understand (partly because of its economic and environmental consequences, which we will discuss shortly). Make *sure* that you understand each step of the derivation and that (in the long run at least), you can reproduce the argument if asked.

T8.5 CONSEQUENCES

As I mentioned before, our technological society extensively employs heat engines in a wide variety of applications. Equation T8.8 therefore has several important conceptual, economic, and environmental consequences.

1. *Any temperature difference can be exploited to generate mechanical energy.* Equation T8.8 tells us that if we have two reservoirs at different temperatures $T_H > T_C$, then in principle we can construct a heat engine that exploits that temperature difference to produce mechanical energy. One of the reasons we use fossil fuels at such a tremendous rate is that it is relatively easy to create a substantial temperature difference by burning such fuels. On the other hand, this fact offers some hope when the fossil fuels eventually run out: *any* means of generating a temperature difference provides an opportunity to convert heat energy to mechanical energy.

2. *The greater the temperature difference, the more efficient the engine.* Equation T8.8 makes clear the advantage in making $T_H - T_C$ as large as possible. Since the most practical low temperature reservoir available to most engines is the surrounding environment (at $T_C \approx 0°C$ to $25°C$), this usually means making T_H as large as possible. The trade-off is usually that constructing engine parts able to withstand extremely high temperatures is expensive, and there comes a point where it becomes more economical to waste fuel than it is to use very expensive materials to boost efficiency by a few percent.

3. *Energy waste is inevitable.* No heat engine is going to be able to convert all of the energy of its fuel into useful mechanical energy: *some* of that energy is necessarily discarded to the cold reservoir. (Indeed, it is the flow of heat energy from hot to cold that drives the heat engine in the first place: heat engines simply take advantage of the fact that such a flow is so strongly driven by the promise of more microstates in the cold reservoir that some of the energy flow can be diverted for useful purposes.) No amount of technological ingenuity will enable one to get around the basic limit on efficiency imposed by the second law of thermodynamics stated by equation T8.8.

This last issue has some serious environmental consequences. Disposing of the inevitable waste heat generated by large heat engines (or large numbers of small engines) in a way that does not detrimentally affect the environment can be a real challenge. Most cities are substantially warmer than the surrounding countryside because of the waste heat being produced within their borders. It has been estimated that the waste heat produced by human activities in the Los Angeles basin now exceeds 7% of the solar energy falling on the basin.

The problem of appropriate disposal of waste heat is particularly acute at large electrical power plants, where very large heat engines are used to convert heat from some source into electrical power. A large electrical plant may need to dispose of several billion joules of waste heat energy into a suitable cold reservoir every second. The environmental consequences of directly dumping this kind of energy into a passing river (the most economical method by far) can be severe: this is not done much any more. Modern cooling towers operate by taking cold river water, spraying it over the pipes containing the working substance to be cooled, and then allowing the river water to evaporate. Since the latent energy of vaporization of water is fairly large, this is an efficient way to carry the waste energy away. While this method does not increase the downstream temperature of the river, it does reduce the amount of water flowing in the river, and this can also have some negative environmental effects (though these are usually less severe than directly dumping the heat in the river).

Exercise T8X.7: In a typical nuclear power plant, heat from the reactor core is used to produce pressurized steam at a temperature of about 300°C (the limit on this temperature is primarily imposed by the dangers of melting the reactor fuel). The temperature of recondensed water leaving the cooling tower is about 40°C. Show that the maximum possible efficiency of the steam turbines in such a plant is about 0.45.

Exercise T8X.8: Due to various unavoidable factors (mechanical friction, the energy needed to pump water around, etc.) the actual efficiency of such a plant is about 0.34. If the plant generates 1000 MW of electrical power (10^9 J/s), show that about 1900 MW of energy has to be given up in the form of waste heat. (This means that nearly two-thirds of the energy in the expensive fuel must be wasted!)

Exercise T8X.9: Imagine that this plant is located on the banks of a major river that is 67 m wide near the plant, an average of 3 m deep, and flows at a rate of about 0.5 m/s. If all of this water could be routed through the plant and the waste heat could be distributed *evenly* throughout this water, show that the river downstream from the plant will be about 4.5°C warmer than the river upstream of the plant. (The specific heat* of water is 4186 J·kg^{-1}·K^{-1}, and the density of water is 1000 kg/m^3.)

Exercise T8X.10: If the water is routed to cooling towers, what fraction of the river has to be evaporated to carry away the waste energy? (The latent energy of vaporization of water is about 2257 kJ/kg. You should find a result of less than 1%: this is going to be much less disruptive of the river's ecology than heating it by 5°C.)

The fact that the efficiency of a heat engine improves as the temperature difference across which it operates gets larger often means that heat engines that exploit natural temperature differences (for example, engines that extract power from oceanic temperature differences, geothermal temperature differences, or temperature differences created using solar energy) are often much less efficient than engines powered by fossil fuels, since the temperature differences available from such sources are typically much smaller than those that can be created by burning fossil fuels. Since heat energy provided by such natural temperature differences is essentially "free", this doesn't matter so much, except that the naturally-driven engines often need to be bigger and more expensive to produce the same amount of power, and the disposal of waste heat can be a more significant problem (see Problem T8S.12 for an example).

This doesn't mean that you cannot extract a *lot* of energy from such sources. A hurricane is essentially a natural heat engine that gets the motive energy for its winds from relatively small temperature differences in the atmosphere. (A hurricane is able to extract lots of energy mainly because it is so big.)

T8.6 REFRIGERATORS

What a refrigerator does

A **refrigerator** is a device that would seem at first glance to violate the second law of thermodynamics. We have seen that heat energy naturally and spontaneously flows from hot to cold because the combined entropy of the hot and cold objects *increases* in this process. The whole purpose of a refrigerator, on the other hand, is to make heat energy flow from cold to hot. How can we do this without violating the second law?

It turns out that we can do this, but we have to *supply* mechanical energy to make this happen. The schematic diagram of a refrigerator presented in Figure

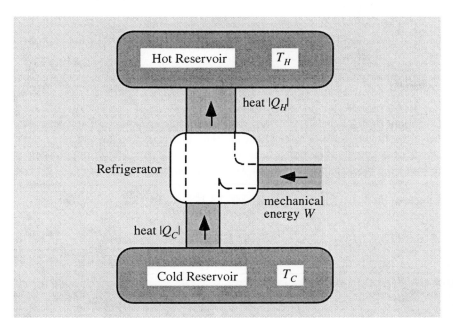

Figure T8.4: A schematic diagram of a refrigerator. Note that a refrigerator is essentially a heat engine operating in reverse: instead of producing mechanical energy from the natural flow of heat from hot to cold, it uses mechanical energy to drive an unnatural flow of heat from cold to hot.

T8.4 makes it clear that a refrigerator is essentially a heat engine operating in reverse (compare with Figure T8.2). Instead of tapping the natural flow of energy from hot to cold to produce a bit of mechanical energy, a refrigerator uses mechanical energy to *drive* the unnatural flow of heat flow from cold to hot!

In the absence of the mechanical energy input, a flow of heat energy from cold to hot certainly *would* violate the Second Law. If energy $|Q_C| = |Q_H|$ were just to flow directly from the cold reservoir to the hot reservoir, the entropy changes of those reservoirs would be:

Second-law restrictions on refrigerator performance

$$\Delta S_C = -\frac{|Q_C|}{T_C}, \qquad \Delta S_H = +\frac{|Q_H|}{T_H} \tag{T8.10}$$

Since $|Q_C| = |Q_H|$ if no mechanical energy is involved and $T_C < T_H$, this implies that $|\Delta S_C| > |\Delta S_H|$, meaning that the *loss* in the entropy of the cold reservoir is greater than the *gain* in the entropy of the hot reservoir. Thus the total entropy of the interacting reservoirs decreases, in violation of the second law.

But we can prevent violation of the second law by adding enough mechanical energy so that $|Q_H|$ becomes enough larger than $|Q_C|$, so that:

$$\frac{|Q_H|}{T_H} \geq \frac{|Q_C|}{T_C}, \quad \text{meaning that } \Delta S_{\text{TOT}} \geq 0 \tag{T8.11}$$

By supplying mechanical energy to the hot reservoir, therefore, we are essentially giving it enough extra entropy to make up for the otherwise larger entropy loss of the cold reservoir.

Exercise T8X.11: Multiply both sides of equation T8.11 by T_H and subtract $|Q_C|$ from both sides to show that

$$|Q_H| - |Q_C| \geq \left(\frac{T_H}{T_C} - 1\right)|Q_C| \tag{T8.12}$$

Note that by conservation of energy, $|Q_H| - |Q_C| = W =$ the mechanical energy that we have to supply. This equation thus specifies how large W must be.

Just as we quantify the effectiveness of a heat engine by describing its efficiency, we can quantify the effectiveness of a refrigerator by specifying its **coefficient of performance** (COP):

Definition of coefficient of performance (COP)

$$\text{COP} \equiv \frac{|Q_C|}{W} \tag{T8.13}$$

We might define a measure of a refrigerator's performance in any number of ways, but this definition talks about the two quantities that we are most interested in: it tells us how much heat we can remove from the cold reservoir in terms of the mechanical energy that we have to put in.

Exercise T8X.12: Combine equations T8.12 and T8.13 to show that

Theoretical maximum COP

$$\text{COP} \leq \frac{T_C}{T_H - T_C} \tag{T8.14}$$

EXAMPLE T8.1

Problem: Estimate the maximum possible coefficient of performance for a standard kitchen refrigerator.

Solution in the case of an ordinary kitchen refrigerator, the cold reservoir (the inside of the refrigerator) has a temperature of about $40°F \approx 5°C \approx 278$ K, while the hot reservoir (the kitchen itself) has a temperature equal to room temperature ≈ 295 K, about 17 K warmer. The maximum theoretical COP for such a refrigerator would be about $(278 \text{ K})/17 \text{ K} = 16.4$. This means that such a refrigerator could (in principle) remove an amount of heat energy from its interior that is more than 16 times larger than the mechanical energy we put in!

Note that COP values will be typically greater than one! We can take advantage of COP greater than one to construct an effective home-heating device.

EXAMPLE T8.2

Problem: A **heat pump** is essentially a refrigerator that moves heat from outside a house (the cold reservoir) to inside the house (the hot reservoir). Imagine that the outside temperature is 0°C and the inside is 22°C, and that the COP of our heat pump is 8.0. **(a)** Show that this COP is possible. **(b)** If the house requires 36 kW of heat energy to keep its temperature at 22°C, how much energy do we have to supply to the heat pump?

Solution **(a)** Since $T_C = 0°C = 273$ K and $T_H = 22°C = 295$ K, the maximum possible COP for the heat pump is $T_C/(T_H - T_C) = (273 \text{ K})/(22 \text{ K}) = 12.4$. Therefore, a COP of 8.0 is certainly possible. **(b)** The problem states that we have to supply an energy $|Q_H| = 36,000$ J to the house every second. Since $|Q_H| - |Q_C|$ is equal to the mechanical energy W that we have to supply, and since equation T8.13 implies that $|Q_C| = (\text{COP})W$, we have

$$W = |Q_H| - |Q_C| = |Q_H| - (\text{COP})W \implies W[1 + \text{COP}] = |Q_H|$$

$$\implies W = \frac{|Q_H|}{1 + \text{COP}} = \frac{36 \text{ kJ}}{9} = 4 \text{ kJ} \tag{T8.15}$$

This is the mechanical energy that we have to supply each second. Therefore, if we use a heat pump, a mere 4 kW of mechanical energy will bring 36 kJ of heat energy into the house, a real bargain!

If everyone would use heat pumps to heat their houses in the winter, a very large amount of energy could be saved. Unfortunately, heat pumps are quite a bit more expensive (and somewhat less reliable) than standard furnaces (it is still relatively inexpensive to heat a home with natural gas, even if supplying 36 kJ of heat requires 36 kJ-worth of natural gas). This makes heat pumps less economically feasible than they should be, considering the value of conserving energy and producing less greenhouse gases.

We will discuss practical designs for both refrigerators and heat engines in the next chapter.

SUMMARY

I. PERFECT HEAT ENGINES
   A. A "perfect" heat engine would convert heat from a reservoir entirely into mechanical energy
      1. But transfers of mechanical energy do not cause entropy changes
      2. So a perfect engine would decrease the reservoir's entropy (by removing heat) with no compensating increase of entropy elsewhere
      3. Therefore, a perfect engine violates the 2nd law of thermodynamics
   B. A classification scheme for perpetual motion machines
      1. "First type" violate the first law (generate energy from nothing)
      2. "Second type" violate the second law (like perfect engine does)

II. REAL HEAT ENGINES
   A. Heat engines take advantage of the natural flow of heat from hot to cold
      1. Such a flow spontaneously occurs because entropy of hot reservoir decreases less than the entropy of the cold reservoir increases
      2. A heat engine converts some of this energy to mechanical energy
      3. But it cannot tap so much that entropy gain in cold reservoir is less than the entropy loss in the hot reservoir.
      4. This strictly limits the amount of energy that can be converted
   B. The efficiency of a heat engine
      1. The efficiency is defined to be $e \equiv W/|Q_H|$ (T8.3)
         where $|Q_H|$ is heat removed from hot reservoir, W is mech. energy
      2. The second law of thermodynamics requires that

$$\Delta S_{TOT} = \Delta S_C + \Delta S_H = \frac{|Q_C|}{T_C} - \frac{|Q_H|}{T_C} \geq 0 \qquad (T8.6)$$

      3. Conservation of energy requires that $W = |Q_H| - |Q_C|$ (T8.2)
      4. After combining these equations and some algebra, we get that

$$e = 1 - \frac{|Q_C|}{|Q_H|} \quad \Rightarrow \quad e \leq 1 - \frac{T_C}{T_H} = \frac{T_H - T_C}{T_H} \qquad (T8.8)$$

   C. Important implications of the discussion above
      1. Any temperature difference can be tapped to produce energy.
      2. Larger temperature difference implies greater efficiency.
      3. Energy waste is inevitable in the case of heat engines.

III. REFRIGERATORS
   A. Definition of a refrigerator: moves heat energy from cold to hot
   B. Essentially a heat engine working in reverse!
      1. instead of tapping natural energy flow to make mechanical energy
      2. uses mechanical energy to drive an unnatural energy flow
   C. How refrigerators avoid violating second law
      1. Normally heat can't flow from cold to hot because entropy loss of cold reservoir would be greater than entropy gain of hot reservoir
      2. But if we add enough mechanical energy to energy going into hot reservoir, we can make its entropy gain larger
   D. The *coefficient of performance* (COP) of a refrigerator
      1. Definition: COP $\equiv |Q_C|/W$ (T8.13)
      2. The limit imposed by the 2nd law: COP $\leq T_C/(T_H - T_C)$ (T8.14)
      3. A refrigerator's COP is commonly greater than one

GLOSSARY

heat engine: a device designed to convert heat energy into mechanical energy. Real heat engines tap into the heat energy that naturally flows from a hot reservoir to a cold reservoir (driven by the second law) and converts part of this flow to mechanical energy.

perpetual motion machine: a hypothetical device that would produce endless amounts of mechanical energy. Perpetual motion machines come in two types, one where the *first* law of thermodynamics would be violated (energy would be created from nothing) and one where the *second* law would be violated (the engine would extract energy from a reservoir in a way that causes the net entropy of the engine and reservoir to decrease).

efficiency e (of a heat engine): defined to be $W/|Q_H|$, that is, the fraction that the mechanical energy W produced by the engine represents of the total heat energy $|Q_H|$ supplied by the engine.

refrigerator: a device designed to use mechanical energy to drive a flow of heat from a cold object to a hot object (opposite to the natural direction of flow).

coefficient of performance COP (of a refrigerator): defined to be $|Q_C|/W$, that is, the amount of heat the refrigerator removes from the cold reservoir expressed as a multiple of the mechanical energy supplied to the refrigerator. COPs are typically greater than 1.

TWO-MINUTE PROBLEMS

T8T.1 Classify the following hypothetical perpetual motion machines as being perpetual motion machines of the first type (A) or the second type (B):

(a) An electric car runs off a battery which drives the front wheels. The car's rear wheels drive a generator which recharges the battery.

(b) An electric car runs off a battery. When the driver wants to slow the car down, instead of applying the brakes he or she throws a switch that connects the wheels to a generator that converts the car's kinetic energy back into energy in the battery.

(c) An engine's tank is filled with water. When the engine operates, it slowly freezes the water in the tank, converting the energy released into mechanical energy.

(d) Compressed air from a tank blows on a windmill. The windmill is connected to a generator that produces electrical energy. Part of that electrical energy is used to compress more air into the tank.

(e) A normal heat engine is used to drive an electrical generator. Part of the power from this generator runs a refrigerator that absorbs the waste heat from the engine and pumps it back into the hot reservoir.

T8T.2 In a maximally efficient heat engine, the amount of entropy that the hot reservoir loses as heat flows out of it is exactly balanced by the entropy that the cold reservoir gains as the waste heat flows into it (T or F).

T8T.3 A heat engine produces 300 W of mechanical power while discarding 1200 W into the environment (its cold reservoir). What is the efficiency of this engine?
A. 0.20
B. 0.25
C. 0.33
D. other (specify)

T8T.4 Imagine that in Iceland, scientists discover geothermal vents that produce abundant pressurized steam with a temperature of 300°C. Engineers construct a heat engine that uses this steam as a hot reservoir and a nearby glacier as a cold reservoir. What is the maximum possible efficiency of this engine?
A. 300%
B. 100%
C. 52%
D. 45%
E. 22%
F. other

T8T.5 Imagine that you are trying to design a personal fan that you wear on your head and operates between your body temperature (37°C) and room temperature (22°C). What is the maximum possible efficiency of this device?
A. 170%
B. 95%
C. 54%
D. 46%
E. 5%
F. other (specify)

T8T.6 Does the COP of a refrigerator get larger or smaller as the temperature difference between its reservoirs increases? A. larger   B. smaller   C. doesn't change

T8T.7 A refrigerator uses 100 W of electrical power and discards 600 W of thermal power into the kitchen. What is its coefficient of performance (COP)?
A. 0.17
B. 0.20
C. 5
D. 6
E. other (specify)
F. impossible: violates conservation of energy

T8T.8 Someone comes to your house selling a device that draws heat from groundwater and supplies that heat to your house. The salesperson claims that the amount of heat energy entering the house will far exceed the electrical energy supplied to the device. What the salesperson claims here is physically impossible (T or F).

HOMEWORK PROBLEMS

BASIC SKILLS

T8B.1 We've seen that heat energy cannot be entirely converted to mechanical energy. Can mechanical energy be entirely converted to heat? If so, give some examples. If not, explain why not.

T8B.2 A heat engine operates between a hot reservoir at 950°C and a cold reservoir at roughly room temperature (22°C). What is its maximum possible efficiency?

T8B.3 An engine uses water at 100°C and 0°C as hot and cold reservoirs. What is its maximum possible efficiency?

T8B.4 A certain heat engine extracts heat energy at a rate of 600 W from a hot reservoir, and discards energy at a rate of 450 W to its cold reservoir. What is its efficiency?

T8B.5 Why is it important that an air conditioner (which is just a kind of refrigerator) be placed in a window or otherwise have access to the hot environment outside? Wouldn't the air conditioner be more effective if it were placed in the center of the room?

T8B.6 Can you cool your kitchen by leaving the refrigerator door open? Can you heat your kitchen by leaving the oven door open? Explain.

T8B.7 A certain air conditioner maintains the inside of a room at 20°C (68°F) when the temperature outside is 37°C (99°F). What is its maximum possible COP?

T8B.8 A refrigerator with a COP of 8.4 uses 300 W of electrical power to extract heat energy from the interior of the refrigerator, which is at 40°F (5°C). At what rate is heat energy removed from its interior?

SYNTHETIC

T8S.1 Is a human being a heat engine? Defend your response.

T8S.2 Imagine that you wish to increase the efficiency of a heat engine. Other things being equal, would it be better to increase the hot reservoir's temperature T_H or decrease the cold reservoir's temperature T_C by the same amount? Explain your response.

T8S.3 Equation T8.9 suggests that an automobile engine should be more efficient in the winter, when the outdoor temperature (and thus T_C) is lower. However, automobiles are actually observed to be *less* efficient in the winter. Speculate as to why this might be.

T8S.4 A hydroelectric power plant, which converts the gravitational potential energy of water stored behind a dam to electrical energy, can operate at very nearly 100% efficiency. How can it do this when a nuclear power plant cannot operate at better than about 40% efficiency?

T8S.5 Imagine that your local power company claims that "heating your home with *electricity* is 100% efficient". In what sense is this true? In what sense is it misleading?

T8S.6 The following wry versions of the first and second laws of thermodynamics have circulated in the physics community for a long time:

> The first law: You can't win.
> The second law: You can't even break even.

Explain in your own words what these restated laws mean.

T8S.7 Do you think that the COP is the most useful way to quantify the effectiveness of a *heat pump* used for home heating? If not, suggest what you think would be a more relevant quantity.

T8S.8 An old friend comes to you asking you to invest in the production of a great new toy: a rubber ball that bounces higher each succeeding bounce (the kinetic energy released during each bounce coming from the internal energy of the rubber). Explain why you should decline this offer.

T8S.9 Current technology permits fossil-fuel burning power plants to produce steam at about 650 K instead of the 570 K that nuclear-powered plants can supply. What is the

maximum possible efficiency of the steam turbines in such a plant?

T8S.10 Your aerospace company wants to construct solar power generators in space. The proposed design involves using gigantic mirrors to concentrate light from the sun on the boiler of a steam turbine. The electrical power generated could be beamed down to Earth using microwaves. Ignoring the problems associated with building the mirrors in space and the difficulties associated with beaming the microwaves, what is a major problem with this proposal?

T8S.11 The temperatures generated in the cylinder of an automobile engine can be in excess of 1500 K. Estimate the theoretical maximum possible efficiency of such an engine. [*Hint*: What is T_C going to be, roughly?]

T8S.12 Imagine a power plant designed to exploit the temperature difference between the ocean surface ($\approx$ 30°C) and the ocean floor ($\approx$ 4°C). Show that a power plant producing 1000 MW of power would have to dump more than 10,000 MW of waste energy into the cold ocean water.

T8S.13 A heat pump has a coefficient of performance of 5.5 and uses 200 W of electrical energy. At what rate does heat energy enter the house if the inside is at 20°C?

T8S.14 An air conditioner with a coefficient of performance of 3.4 uses 1000 W of electrical power. At what rate does it extract heat energy from the house?

T8S.15 The inside temperature of a house is 23°C. The outside temperature is –5°C. A heat pump is used to heat the house. It is able to move heat energy into the house at a rate of 1200 W while using an average of 300 W of electrical energy. What is the real coefficient of performance of this heat pump?

T8S.16 A freezer with a coefficient of performance of 4.9 uses 250 W of power. How long would it take such a freezer to freeze 15 kg of water?

T8S.17 Imagine that a site in Iceland is discovered where temperatures a relatively short distance below the surface of the earth are in the range of 600°C (due to the relatively close proximity of molten rock upwelling from deep inside the earth). A geothermal power plant is constructed at this site. The cold reservoir for this plant is a pool of water constantly fed with ice from a glacier. If the plant produces 100 MW of electrical energy, what is the rate at which ice is melted?

T8S.18 Imagine a solar electrical power plant that operates using a gigantic mirror to concentrate light on the boiler for a steam engine. Assume that the boiler generates steam at a temperature of 550°C. The only possibility for a cold-temperature reservoir is a nearby creek that has an average width of 4 m and an average depth of 0.7 m, flowing at a rate of 0.7 m/s. What is the maximum electrical power that could be produced by the plant if it boils the entire creek dry?

T8S.19 Let's use a heat pump to cook dinner! Say that the temperature of your kitchen is 23°C. You set up a heat pump that has 45% of its theoretical maximum COP, and you want to be able to supply 700 W of heat energy for boiling your soup. How much electrical energy do you have to supply to the heat pump? How much better is this (if any) than simply using an electric hot plate?

RICH-CONTEXT

T8R.1 Is it more economical to use a heat pump to heat your home during the winter in California or in Minnesota? Assume that energy costs and house designs are the same in both locations. Also assume that the rate at which a house loses energy is proportional to the temperature difference between the outside and inside of the house.

T8R.2 Imagine that you are the CEO of a power company building a new power plant. Imagine that you have a choice of using coal or oil. Imagine that the cost of coal is projected over the decade to be about 75% times that of oil needed to produce the same thermal energy. On the other hand, the contractor proposing the oil-fired design promises to deliver steam at 350°C, while the contractor pro-posing a design based on coal can only promise steam at 290°C. It also costs about 15% more per MJ of thermal energy produced to clean up emissions from coal burning than to clean emissions from oil burning. Other things being equal, which is the more economical choice?

ADVANCED

T8A.1 Heat pump performance might better be quantified by the quantity $h \equiv |Q_H|/W$. Why might this be a more useful quantity for describing a heat pump than describing a refrigerator? Derive an expression for limits imposed on this quantity by the second law of thermodynamics. Carefully explain your derivation.

ANSWERS TO EXERCISES

T8X.1 The entropy of the hot reservoir decreases as the engine extracts energy from it, but the engine causes no associated increase in the entropy of anything else (the mechanical energy released does not necessarily change the entropy of anything else.) Therefore the "isolated system" consisting of the reservoir and engine suffers a decrease in entropy in this process, contrary to the second law.

T8X.2 As figure T8.2 shows, the engine splits the energy $|Q_H|$ extracted from the hot reservoir into two parts: mechanical energy W and heat energy $|Q_C|$ going to the cold reservoir. Conservation of energy implies that $W + |Q_C| = |Q_H|$ or $|Q_H| - |Q_C| = W$.

T8X.3. Substituting $W = |Q_H| - |Q_C|$ into equation T8.3

$$\Rightarrow \quad e = \frac{|Q_H| - |Q_C|}{|Q_H|} = 1 - \frac{|Q_C|}{|Q_H|} \qquad (\text{T8.16})$$

T8X.4 The second law of thermodynamics requires that the total entropy of the interacting objects here (the two reservoirs and the engine) must not decrease:

$$\Delta S_{\text{TOT}} = \Delta S_C + \Delta S_{\text{engine}} + \Delta S_H \geq 0 \qquad (\text{T8.17})$$

As discussed, $\Delta S_{\text{engine}} = 0$ (at least in the long run). Substituting the values of ΔS_H and ΔS_C from equation T8.5 yields the desired result.

T8X.5 Multiplying both sides by $T_C|Q_H|$, we get

$$\frac{|Q_C|}{|Q_H|} - \frac{T_C}{T_H} \geq 0 \qquad (\text{T8.18})$$

Adding T_C/T_H to both sides gives the desired result.

T8X.6 Equation T8.4 reads

$$e = 1 - \frac{|Q_C|}{|Q_H|} \qquad (\text{T8.19})$$

This will be *smaller* than the quantity $1 - T_C/T_H$, because according to equation T8.7, $|Q_C|/|Q_H| \geq T_C/T_H$, so subtracting $|Q_C|/|Q_H|$ from 1 yields a smaller number than subtracting T_C/T_H from 1. Therefore

$$e \leq 1 - \frac{T_C}{T_H} \qquad (\text{T8.20})$$

as stated.

T8X.7 $e \leq (T_H - T_C)/T_H = (573\text{ K} - 313\text{ K})/573\text{ K} = 0.45$.

T8X.8 Every second, the plant produces $W = 1000$ MJ of mechanical energy. If the efficiency is 0.34, this means that $|Q_H| = W/e = (1000\text{ MJ})/0.34 = 2900$ MJ. This means that $|Q_C| = |Q_H| - W = 1900$ MJ. This is the energy budget every second, so the rate of waste heat flow is 1900 W.

T8X.9 The flow rate of the river is $(67\text{ m})(3\text{ m})(0.5\text{ m/s}) = 100\text{ m}^3/\text{s} = 10^5$ kg/s. In one second, then, the 1900 MJ of waste heat produced by the plant has to be distributed in 10^5 kg of water. The resulting increase in temperature is

$$dT = \frac{dU}{mc} = \frac{1.9 \times 10^9\text{ J}}{(10^5\text{ kg})(4186\text{ J}\cdot\text{kg}^{-1}\cdot\text{K}^{-1})} = 4.5\text{ K} \quad (\text{T8.21})$$

T8X.10 The energy required to vaporize a mass m of water is $dU = mL$, where L is the latent heat* of vaporization. The amount of water vaporized by 1900 MJ of energy is thus

$$m = \frac{dU}{L} = \frac{1.9 \times 10^9\text{ J}}{2.257 \times 10^6\text{ J/kg}} = 840\text{ kg} \qquad (\text{T8.22})$$

This is less than a cubic meter of water, so less than 1% of the river would have to be vaporized.

T8X.11 The answer is given (follow the instructions).

T8X.12 Since $W = |Q_H| - |Q_C|$, we have

$$\text{COP} = \frac{|Q_C|}{W} = \frac{|Q_C|}{|Q_H| - |Q_C|} \qquad (\text{T8.23})$$

Plugging equation T8.12 into this (and noting that when we *divide* by a bigger number the result is *smaller*) we get

$$\text{COP} \leq \frac{|Q_C|}{(T_H/T_C - 1)|Q_C|} = \left(\frac{T_H}{T_C} - 1\right)^{-1} \qquad (\text{T8.24})$$

Putting the quantity in parentheses over a common denominator and then inverting yields equation T8.14.

T9

PRACTICAL ENGINES AND REFRIGERATORS

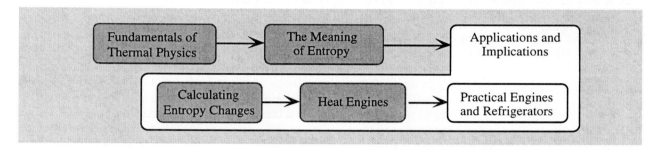

T9.1 OVERVIEW

In the last chapter, we discussed heat engines and refrigerators in abstract and how the second law of thermodynamics limits the efficiency of such engines. In this chapter, we will draw on the knowledge about the behavior of ideal gases that we developed in chapter T3 to discuss several different designs for actual heat engines and refrigerators.

Here is an overview of the sections in this chapter.

T9.2 *THE CARNOT CYCLE* discusses the Carnot engine, which is the idealized archetype for engines based on ideal-gas processes.

T9.3 *THE STIRLING ENGINE* explores the very clever Stirling engine, the most practical external combustion engine using an ideal gas as a working substance.

T9.4 *THE AUTOMOBILE ENGINE* examines the basic principles behind the automobile engine, the heat engine most commonly used today.

T9.5 *REFRIGERATORS* discusses the most commonly-used designs for household refrigerators.

T9.6 *THERMOELECTRIC DEVICES* discusses the phenomenon of thermoelectricity and how we can use this effect to construct thermoelectric generators and refrigerators. This section illustrates that the same laws of thermodynamics apply even to devices that do *not* use gases as a working substance.

T9.7 *CLOSING COMMENTS* takes a retrospective look at the unit and the complex relationship between physics, technology, and society.

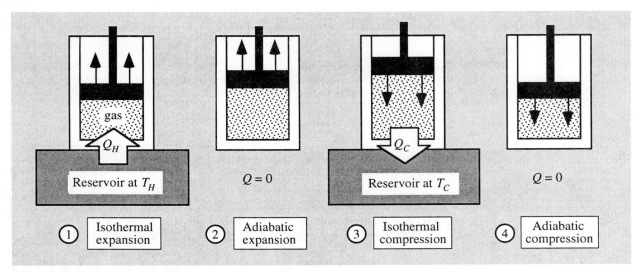

Figure T9.1: The Carnot cycle. You should imagine that the piston is connected to a flywheel (see Figure T9.2a) that moves the piston in and out and stores the mechanical energy produced by the cycle.

Steps in the Carnot cycle

T9.2   THE CARNOT CYCLE

The **Carnot cycle**, which was first described by the French physicist Sadi Carnot (pronounced *car-noh*) in 1824, is a hypothetical cyclic process that can be used to convert heat to mechanical energy using an ideal gas as the working substance. While this particular cycle is not used in realistic engines (for a variety of reasons), it does describe an idealized sequence of gas processes that could be used in principle to produce mechanical energy at the maximum theoretical efficiency allowed by the second law of thermodynamics. Understanding the Carnot cycle also prepares us for considering more realistic engines later in the chapter.

The Carnot cycle uses an ideal gas confined in a cylinder by a piston, as shown in Figure T9.1. A massive flywheel (see Figure T9.2a) keeps the piston moving in and out of the cylinder at a regular pace and stores the mechanical energy produced by the engine. The Carnot cycle consists of the four steps shown in Figure T9.1. A detailed description of the steps follows.

(1) Just as the flywheel begins to pull the piston out, we put the cylinder in thermal contact with a reservoir at temperature T_H. This keeps that gas temperature fixed at T_H. Since the thermal energy U of an ideal gas depends on N and T but not V, this means that U also is fixed. However, the expanding gas does work on the piston, so to keep U fixed, heat must flow into the gas from the reservoir. Let this amount of heat be $|Q_H|$.

(2) In the second step, we remove the reservoir. The gas expands adiabatically as the piston continues to move out. Work energy continues to flow out of the gas as it expands, but no heat is coming in to replace it, so the thermal energy of the gas decreases, and thus its temperature decreases. The expansion continues until the gas temperature falls to T_C and the flywheel has pulled the piston to its maximum outward position.

(3) In the third step, the gas is gently compressed (as the flywheel now begins to push the piston *into* the cylinder) while it is in contact with the cold reservoir at temperature T_C. Because the gas is being compressed, work energy flows *into* the gas. Since its temperature (and thus its thermal energy) does *not* increase, an equal amount of heat energy must flow out of the gas into the cold reservoir. Let's call this amount of heat $|Q_C|$.

(4) In the final step, the reservoir is removed. The piston continues to compress the gas (now adiabatically). Now work energy flows into the gas, but no heat flows out, so the thermal energy and temperature of the gas increase. The compression continues until the gas temperature reaches T_H, and the piston has moved to its innermost position. The gas is now in the same state as when the cycle started.

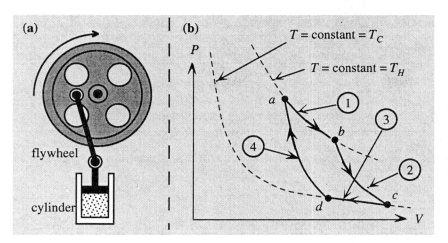

Figure T9.2: (a) This drawing shows how the flywheel is attached to the piston in such a way that it moves the piston in and out of the cylinder as it turns. **(b)** A PV diagram of the Carnot cycle process. The net work energy flowing out of the gas in the cycle is the area enclosed by the process on the diagram.

In a certain sense, the gas is used in this process to carry heat from the hot reservoir to the cold reservoir. In this process, however, there is a net flow of mechanical energy out of the gas: the gas converts *some* of the heat it carries to the cold reservoir to mechanical energy. How do we know? Consider the PV diagram shown in Figure T9.2b. The two curves of constant $T = T_H$ and $T = T_C$ are shown as dotted curves. These are curves where $P \propto 1/V$, since the ideal gas law implies that when the temperature is constant, PV = constant. The relationship between P and V during an adiabatic process is given by equation T3.13a:

Why this cycle produces net mechanical energy

$$PV^\gamma = \text{constant}, \quad \gamma = 1 + 2/n \qquad \text{(T9.1)}$$

where n is the number of active degrees of freedom available to each gas molecule ($n = 3$ for a monatomic gas and 5 for a diatomic gas). This equation implies that $P \propto V^{-\gamma}$, and since $\gamma > 1$, this means that P falls off more sharply with increasing volume than the curves for the isothermal process, as shown.

Remember that the work done on or by a gas in a given quasistatic process is equal to the area under the curve representing that process on a PV diagram. The diagram above clearly shows that the area under the curves of processes 1 and 2 is greater in magnitude than the area under the curves of processes 3 and 4. The mechanical energy moving *out* of the gas during the expansion processes is thus greater than the mechanical energy moving into the gas during the compression processes. Thus, during the cycle as a whole, mechanical energy leaves the gas (and is stored in the flywheel). This must come at the expense of some part of the heat energy that it absorbed in the first step.

One of the main reasons that the Carnot cycle is of interest is that such an engine (in the absence of friction or other imperfections) would operate at the maximum efficiency allowed by the second law. Let's see if we can show this.

The efficiency of a Carnot engine

Consider the isothermal expansion in Step 1. Because the temperature of the gas remains constant, work energy that flows out of the gas as it expands must be balanced by the heat energy flowing into the gas:

$$|Q_H| = |W_1| \qquad \text{(T9.2a)}$$

According to equation T3.10b, the work involved in an isothermal process is:

$$W_1 = -Nk_B T_H \ln(V_b/V_a) \qquad \text{(T9.2b)}$$

where V_a is the volume of the gas at the point marked a and V_b is the volume of the gas at the point marked b in Figure T9.2b. Since the gas is expanding, work is flowing *out* of the gas, meaning that W_1 is negative. This means that $|W_1| = -W_1$, and therefore that

$$|Q_H| = + Nk_BT_H \ln(V_b/V_a) \qquad (T9.2c)$$

Exercise T9X.1: In a similar manner, show that for step 3, we have

$$|Q_C| = +Nk_BT_C \ln(V_c/V_d) \qquad (T9.3)$$

(*Hint*: be very careful with signs, and remember that $\ln(1/x) = -\ln x$.)

If we divide equation T9.3 by T9.2b, we get:

$$\frac{|Q_C|}{|Q_H|} = \frac{T_C \ln(V_c/V_d)}{T_H \ln(V_b/V_a)} \qquad (T9.4)$$

Now let us consider the adiabatic processes. According to equation T9.1 and the ideal gas law, we have:

$$\text{constant} = PV^\gamma = (PV)V^{\gamma-1} = (Nk_BT)V^{\gamma-1} \qquad (T9.5a)$$

Absorbing Nk_B into the constant, we have

$$\text{constant} = TV^{\gamma-1} \qquad \text{(for adiabatic processes)} \qquad (T9.5b)$$

Applying this to the situation at hand, this means that

$$T_H V_b^{\gamma-1} = T_C V_c^{\gamma-1} \quad \text{and} \quad T_H V_a^{\gamma-1} = T_C V_d^{\gamma-1} \qquad (T9.5c)$$

Exercise T9X.2: Show that this implies that

$$\frac{V_c}{V_d} = \frac{V_b}{V_a} \qquad (T9.6)$$

Exercise T9X.3: Finally, combine this with equation T9.4 to show that

$$\frac{|Q_C|}{|Q_H|} = \frac{T_C}{T_H} \qquad (T9.7)$$

Since the efficiency of a heat engine is defined to be $e = W_{\text{net}}/|Q_H|$, and the net mechanical energy produced has to be $W_{\text{net}} = |Q_H|-|Q_C|$, we have in this case

$$e = \frac{|Q_H|-|Q_C|}{|Q_H|} = 1-\frac{|Q_C|}{|Q_H|} = 1-\frac{T_C}{T_H} = \frac{T_H-T_C}{T_H} \qquad (T9.8)$$

Thus the efficiency of a Carnot engine (which we have calculated here without reference to the second law) happens to be the same as the maximum possible efficiency that *any* heat engine can have (see equation T8.9), according to the limits imposed by the second law.

T9.3 THE STIRLING ENGINE

In 1816, the Scottish clergyman Robert Stirling developed a practical heat engine using air as a working substance. Stirling engines were used in a variety of applications throughout the 19th century before they were gradually displaced by internal combustion engines. The **Stirling engine** does have some interesting features that make it still a subject of research today. Its design is simple and involves few moving parts. The heat source is external to the engine (instead of

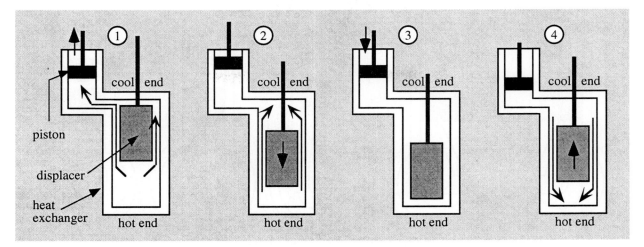

Figure T9.3: A schematic diagram of a Stirling engine.

relying on combustion inside the engine, as in the case of an automobile engine). This means that the Stirling engine can use any kind of heat source (including the focused light of the sun). There has been some interest in Stirling engines recently because of the fact that fossil fuels can be burned much more cleanly and efficiently in open air than inside an internal combustion engine.

How a Stirling engine operates

The operation of a Stirling engine is illustrated in Figure T9.3. The piston and the displacer are both connected to a flywheel at points 90° apart, so that when the piston is at rest the displacer is moving and vice versa. The Stirling cycle, like the Carnot cycle, has four steps. In the first step, air at the hot end of the heat exchanger expands, pushing around the displacer and against the piston, driving it out. In the second step, the piston comes essentially to rest and the displacer moves toward the hot end of the exchanger, displacing the air to the cold end of the exchanger. In the third step, the flywheel begins to push the piston in, but now most of the air inside the engine is at the cool end of the heat exchanger, and so resists the compression less. Finally, the piston comes to rest again and the displacer moves toward the cold end, displacing the air back to the hot end, and we are now back where we started.

Advantages and disadvantages in practical cases

The design of the Stirling engine is masterful in its simplicity and reliability, but there are several things that make it less than ideal in operation. If the engine cycles rapidly, the air doesn't have much time at either end of the exchanger to either heat up fully or cool down completely. This means that the Stirling engine will not take full advantage of the available temperature difference.

The main competition against Stirling engines in the 1800's was the steam engine. Steam engines were much more complex and dangerous than Stirling engines, but they had one tremendous advantage: a smaller steam engine could produce much more power. In a steam engine, boiling water produced high pressure steam that pushed against a piston. When water undergoes a phase change, it increases its volume a thousand-fold under ordinary pressures. What this means is that steam engines can generate much larger pressures with a given heat source than any Stirling engine could, meaning that much more demanding loads could be handled by steam engines.

The Stirling engine was finally completely crowded out by the gasoline internal combustion engine in the early part of this century. In a gasoline engine, the fuel explodes inside the cylinder of the automobile, again creating impressive pressures. The Stirling engine is simply not capable of generating large pressures: to produce a significant amount of work energy, it must cycle rapidly, so that the small amount of work that it does each cycle at least is produced at a greater rate. Unfortunately, the Stirling engine is less efficient when the cycles are rapid, because the air heats and cools imperfectly.

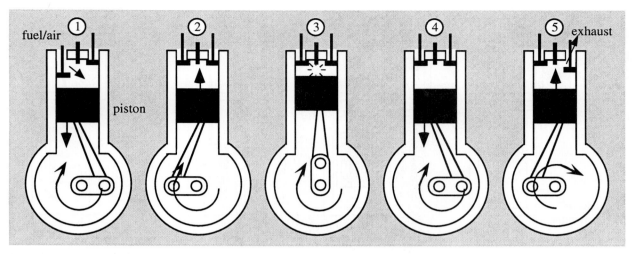

Figure T9.4: The five steps in an automobile engine cycle.

T9.4 THE AUTOMOBILE ENGINE

Steps in the automobile engine cycle

The operation of an idealized automobile engine is illustrated in Figure T9.4. We can consider each cycle of the engine to be divided into five steps.

(1) During the first step, the piston moves down the cylinder, drawing in mixture of gasoline and air through a valve that is opened at the top of the cylinder. This step is called the **intake stroke**.

(2) During the second step, the valve is closed and the piston moves back up the cylinder, compressing the gasoline/air mixture roughly adiabatically. This step is called the **compression stroke**.

(3) When the gas is fully compressed, a spark from the spark plug at the top of the cylinder causes the gasoline/air mixture to explode. (In a diesel engine, the increase in temperature caused by the adiabatic compression itself ignites the gas: no spark plug is necessary.)

(4) The hot gases produced by the explosion drive the piston down (this is the **power stroke**). The exhaust gases expand adiabatically during this step.

(5) Finally, a valve is opened and the upward-moving piston pushes the exhaust gases out of the cylinder (this is the **exhaust stroke**). This returns the engine to its original state.

Note that the piston has to go through two complete cycles of its motion (four "strokes") to go once through these five steps. Such an engine is called a **four-stroke** gasoline engine. The smaller engines used in lawnmowers and other such devices are **two-stroke** engines: such engines combine the exhaust and intake steps when the piston is lowest just after the power stroke so that the piston only has to go around once to get through these five steps. Two-stroke engines are often simpler and lighter than four-stroke engines but not as efficient or clean (fuel and air usually gets mixed with exhaust in the process).

Discussion of *PV* diagram

Figure T9.5 shows an idealized *PV* diagram for an automobile engine. During the intake stroke (Step 1), the volume of the cylinder changes from V_a to V_b, drawing in the gasoline/air mixture at atmospheric pressure and ambient temperature. At point b, the valve closes and the gas is adiabatically compressed to volume V_a again (point c): this is the compression stroke. The gasoline is then ignited, and the temperature of the gas (and thus its pressure) suddenly increases while the piston is almost at rest. The power stroke (Step 4) involves an adiabatic expansion of the exhaust gases back to volume V_b. At point e, the exhaust valve is opened, and the pressure in the cylinder suddenly drops to atmospheric pressure, taking the exhaust gas back to point b again. (The explosive decompression during this step is why gasoline engines are noisy and require mufflers.) Finally, the exhaust gas is pushed out of the cylinder at atmospheric pressure during the exhaust stroke (Step 5).

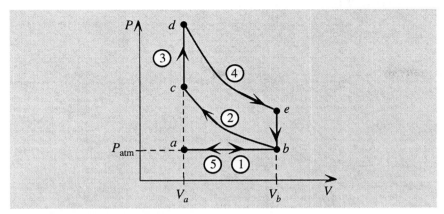

Figure T9.5: An idealized PV diagram for an automobile engine cycle. Step $e \to b$ happens when the exhaust valve opens and the exhaust gas expands freely into the atmosphere. Note also that during steps (1) and (5) the number of molecules in the cylinder is not constant.

(Some caution is advised in interpreting this diagram, as the amount of gas in the cylinder changes in steps 1, 5 and in the explosive decompression $e \to b$: we are accustomed to interpreting PV diagrams assuming that N is constant. For example, during Step 5, we see the pressure remain constant while the volume decreases. If a *fixed* amount of gas were involved, this would imply that the temperature of the gas decreases in the process (since $PV = Nk_BT$). In this case though, N drops and T remains about the same.)

You can use the information on this diagram to determine the efficiency of such a gasoline engine (see Problem T9S.2). The efficiency turns out to be:

Efficiency of the automobile engine

$$e \equiv \frac{|W_{net}|}{|Q_H|} = 1 - \frac{1}{(V_b / V_a)^{\gamma-1}} \qquad (T9.9)$$

Note that the efficiency depends *only* on the compression ratio V_b/V_a. A typical automobile compression ratio might be $V_b/V_a = 8$: such a ratio implies a theoretical engine efficiency of about 0.56. The actual efficiency (due to effects of friction, incomplete burning of the fuel and other imperfections) is more like 0.20, meaning that an automobile engine typically throws away 4/5 of the energy available in the gasoline!

Diesel engines are constructed to have a higher compression ratio (≈ 16), which makes them somewhat more efficient. On the other hand, the large pressures generated by such compression ratios mean that diesel engines have to be thicker and heavier than automobile engines. Thus diesel engines are generally more suitable for trucks and other large vehicles than for automobiles.

Exercise T9X.4: Use equation T9.9 to verify that the theoretical efficiency of an automobile with a compression ratio of 8 is indeed about 0.56. (The value of γ for air is 7/5 since air is composed almost entirely of diatomic molecules.)

Exercise T9X.5: Show that equation T9.5b implies that

$$\frac{T_c}{T_b} = \frac{T_d}{T_e} = \left(\frac{V_b}{V_a}\right)^{\gamma-1} \qquad (T9.10)$$

Exercise T9X.6: Show that this means that the efficiency of the automobile engine can be written

$$e = 1 - \frac{T_b}{T_c} \qquad (T9.11)$$

and that this is *smaller* than the maximum efficiency allowed by the second law.

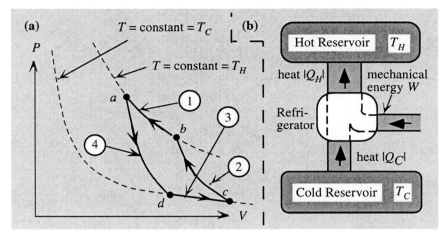

Figure T9.6: (a) A *PV* diagram for a Carnot engine that we force to run in reverse. **(b)** The net effect of the cycle is to extract heat energy from the cold reservoir and deposit heat into the hot reservoir.

T9.5 REFRIGERATORS

Reversed Carnot engine acts like a refrigerator

You can create a refrigerator by running a Carnot engine backwards, that is, by *compressing* the gas during steps (1) and (2) and expanding it in steps (3) and (4) of the cycle we discussed in section T9.2. Figure T9.6 shows a *PV* diagram of the resulting process. Note that the area under the curves for the compression processes (1) and (2) is greater than that for the expansion processes (3) and (4). Since work energy flows into the gas in the compression process, this means that more mechanical energy flows into the gas during the cycle than flows out: we have to put mechanical energy *into* the Carnot engine to run it in reverse.

Since the gas returns to the same state (and thus the same thermal energy) at the end of each cycle, this means that the mechanical energy that we put in has to flow out in the form of heat during some part of the cycle. This cannot happen during steps 2 and 4, because these steps are adiabatic. During step 3, the gas is expanding while it is in contact with the cold reservoir. This means that work energy flows *out* of the gas, but since it is being held at a constant temperature (and thus constant thermal energy), heat energy $|Q_C|$ must be flowing *into* the gas from the cold reservoir. During step 1, on the other hand, the gas is being compressed while in contact with the hot reservoir. Thus work energy flows *into* the gas. Since the gas is being held at a constant temperature (and thus constant thermal energy), heat energy must flow *out* into the hot reservoir during this process. We see, therefore, that this reversed Carnot cycle extracts heat from the cold reservoir and deposits it (along with the mechanical energy we supply) in the hot reservoir. This is exactly what a refrigerator should do!

A reversed Stirling engine would also work as a refrigerator. However, it is difficult to build a practical refrigerator using either of these technologies (such a refrigerator would be too bulky to be economical).

Modern design for a refrigerator using phase change

Most modern refrigerators instead use the process shown in the (simplified) diagram shown in Figure T9.7. A liquid with suitable characteristics circulates through the system shown, driven by a **compressor** (a pump). The compressor pushes the liquid through the **condenser** pipe at a fairly high pressure. The liquid then sprays through a narrow **expansion valve** into a length of pipe inside the refrigerator called the **evaporator**. The pressure inside the evaporator is kept low by the compressor pump: the harder the pump works, the lower the pressure in the evaporator will be. The boiling point of a liquid generally depends strongly on the surrounding pressure, so by appropriate choice of liquid and low pressure in the evaporator, the liquid can be induced to vaporize in the evaporator at the temperature inside the refrigerator. As the liquid evaporates, it absorbs the heat required for its phase change from the inside of the refrigerator. After absorbing this heat, the vapor is pumped by the compressor into the condenser piping, where the high pressure produced by the pump causes the vapor to

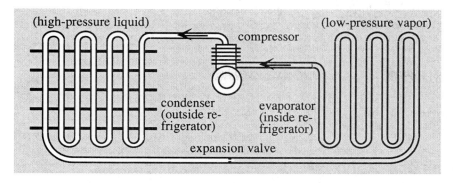

Figure T9.7: Schematic diagram of a modern refrigerator.

condense, releasing the heat that it absorbed in the refrigerator interior (plus the mechanical work applied by the pump) in the condenser piping. This piping is placed outside the refrigerator (often on its back) so that the heat released can be carried away by the air in the kitchen.

This scheme is better than using a reversed Carnot cycle because *much* more energy can be absorbed per unit mass of material undergoing a phase change than by any gas process. Thus, a vapor-compression refrigerator can be much smaller and lighter than a Carnot refrigerator could be.

The choice of working liquid is clearly important: the liquid must have a boiling point that varies strongly with pressure and is in the appropriate temperature range. Until quite recently, most refrigerators were constructed using Freon (an ozone-damaging chlorofluorocarbon) as a working substance, but manufacturers are currently shifting to fluids that do less damage to the earth's ozone layer.

T9.6 THERMOELECTRIC DEVICES

The lowest possible energy level for conduction electrons in a metal or semiconductor has a value that is different in different materials. We can consider this energy level to be like an internal potential energy V for these electrons due to their internal interactions with the conducting material.

Theory of operation of a thermoelectric generator

When two different materials are placed in electrical contact, electrons will spontaneously move to the material where they have the lower V, because this is where the microstates are: the more energy that electrons collectively have above this basic ground level the more ways there are to distribute that energy among them. Thus if two different materials A and B are connected in a circuit, as shown in Figure T9.8a, electrons will spontaneously leave material A (assuming that it has the higher V) and enter B until B has such a large negative charge as to effectively erase the energy advantage of going to B. However, there is no advantage for an electron to go *around* the circuit: the entropy gained by an electron leaving A on the right is the same as the entropy lost by an electron entering A on the left. Since there is no net entropy gain for an electron going around the circuit, no current will spontaneously flow.

This assumes, however, that the two junctions are held at the same temperature. However, the value of V for a substance also depends on temperature, and may vary in different ways for different materials. Imagine for the sake of simplicity that V_B is essentially independent of temperature, while V_A decreases somewhat with increasing temperature. Therefore, if the two junctions are held at different temperatures, the entropy gain implied by the big jump in V between A and B on the right drives electrons across this junction. This generally tends to electrostatically push electrons around the circuit. Electrons pushed across the junction between B and A on the right will lose entropy as they cross the left junction, but not as much as they gain on the right. This means that the system experiences a net entropy gain when electrons move clockwise around the circuit: therefore, electrons will spontaneously flow in this direction.

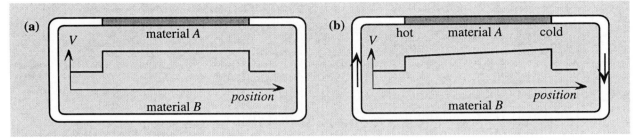

Figure T9.8: (a) A circuit consisting of two different types of materials wired in series. The graph in the middle shows the effective potential energy V of an electron in various parts of the circuit. **(b)** If V_A varies with temperature while V_B does not, we can create a bigger potential energy drop at the right junction than at the left junction if we keep the junctions at different temperatures. This gives rise to an entropy gain if electrons flow around the circuit.

Practical applications

Of course, energy must come into the system from *somewhere* to replace the energy lost by the electrons as they flow around the circuit. Note that an electron has to absorb energy to climb the potential energy barrier on the left: it gets this energy from thermal energy at the hot right junction. As the electron goes over the potential energy drop on the right, it converts some of its potential energy to thermal energy (which is absorbed by whatever is keeping the junction cold) and some (by building up a negative charge just to the left of the junction) to electrostatic potential energy (which the electron eventually loses by going around the circuit).

So the flowing electrons essentially transport thermal energy from the hot junction to the cold junction, converting some of that energy to electrical energy. This means that the device illustrated in Figure T9.8b is a *heat engine* that uses electrons as the working substance and produces electrical energy directly.

The differences in V between most materials are so tiny that this effect is hard to observe. However, certain kinds of materials, notably semiconductors made of bismuth telluride (Bi_2Te_3) have the right characteristics to construct a practical **thermoelectric generator**. Such generators are simple in construction, are completely silent, and have no moving parts. Unfortunately, the efficiency of such devices for practical temperature differences tends to be on the order of 5% (much lower than the efficiency allowed by the second law), so thermoelectric generators are not economical for large-scale power generation.

However, thermoelectric generators are useful in cases where small size, silent operation, and reliability are essential. One can construct a compact generator that converts nuclear energy to electric energy by sealing a radioactive substance inside a case and using the heat generated as the substance decays as a hot reservoir for a thermoelectric generator. Such generators are commonly used to provide power for flashing lights in inaccessible lighthouses, for automatic weather stations at the north and south poles, and for satellites and interplanetary probes. Thermoelectric devices are also often used in gas home furnaces as a safety feature. A thermoelectric device uses heat from the pilot flame to generate electricity that keeps the main gas valve open. If the pilot flame ever goes out, the valve will then shut automatically, keeping gas from escaping.

Thermoelectric refrigerator

By forcing current through the thermoelectric generator in the reverse direction, one can force the electrons to draw energy in from the cold junction and release it at the hot junction, making the generator into a refrigerator. It is possible to purchase small thermoelectric refrigerators for use in cars and boats, but such devices are fairly expensive. Thermoelectric refrigerators are more commonly used in scientific and medical applications where very small, reliable, and/or silent refrigerators are needed (for transporting tissue samples, for example).

T9.7 CLOSING COMMENTS

Why are some physical processes irreversible? How does this irreversible behavior arise out of the reversible laws of microscopic physics? We have come a long way in this unit from our initial statement of these puzzling questions.

We have discovered two important things in our journey that I'd like to highlight in closing this unit.

First of all, an answer to these questions seemed *impossible* at first glance (how could reversible microscopic behaviors possibly lead to irreversible macroscopic behavior). Yet the answer that we eventually discovered was simple, beautiful, and (in retrospect) very straightforward. What it required was only a shift of vision, being able to look at the problem in a new way. The keys we needed were some careful thinking about the relationship between macrostates and microstates and the recognition that what looks like *certain* irreversible behavior on the macroscopic scale may in fact only be *very probable* behavior.

Change of mental perspective crucial for pushing physics forward

This is part of the wonder and beauty of physics, to see how a small shift of mental gears can open up whole new vistas of insight into the way the universe works. The other continually surprising thing is that these ideas *work* and have useful implications. In the last two chapters especially, we've seen that the concept of entropy has important technological applications and societal implications. Entropy illuminates how heat engines work and how they don't work. The second law of thermodynamics not only prevents us from wasting time pursuing fruitless projects (such as perpetual motion machines) but can also direct our attention to new and clever ways to take *advantage* of this law to do useful things.

The quest for simplicity, coherence, and beauty: these are the things that physicists cherish and strive to achieve. Our larger society tolerates physicists' devotion to such goals because it often turns out that knowing about the universe also leads to new and useful applications of that knowledge. There is, however, an inherent tension between the goals of communion with the universe and the production of gadgets to feed our need for economic progress, and this tension can be destructive of both physics and society. The history of thermal physics, like the history of all of physics, embraces both poles of this never-ending dialog: the joy of unexpected coherence, the host of applications on which our civilization is based, and the societal and environmental problems that come with those applications.

The tension between knowledge for its own sake and practical applications

I hope that this unit (and indeed the course in general) has increased your understanding and appreciation of physics as a triumph of the human spirit and imagination, as a means of developing closer communion with the world in which we live, and as a source of practical knowledge that (*if* coupled with sound judgment and moral vision) can and should (like all true knowledge) provide a higher quality of life for our society as a whole.

SUMMARY

I. THE CARNOT CYCLE
- A. The Carnot cycle is a sequence of ideal gas processes that converts heat to work. It is not usually very practical, but it serves as an archetype of an idealized heat engine.
- B. The Carnot cycle consists of four steps
  1. isothermal expansion in contact with a hot reservoir at $T = T_H$
  2. adiabatic expansion to temperature T_C
  3. isothermal compression in contact with a cold reservoir at $T = T_C$
  4. adiabatic compression back to temperature T_H.
- C. This cycle
  1. absorbs heat during step 1 (to keep $U \propto T$ constant as gas expands)
  2. rejects heat during step 3 (to keep U constant as gas is compressed)
  3. produces a net amount of mechanical energy
- D. The efficiency of this cycle is equal to the second-law limit

II. THE STIRLING ENGINE
- A. This is a simple, practical heat engine using a gas as working substance
- B. The basic process:
  1. it uses a long cylinder having a hot end and a cool end
  2. a displacer shuffles air back and forth between its hot and cool ends
  3. this is timed so that the air is heated as piston moves out, cooled as it moves in
  4. this produces net outgoing mechanical energy in cycle
- C. It is generally too large and inefficient to be economical

III. THE AUTOMOBILE ENGINE
- A. Generally more economical than ideal-gas engines because even a small engine can generate lots of pressure (and thus power)
- B. *Four-stroke* automobile engine cycle has five distinct steps:
  1. *intake stroke*, where outgoing piston pulls in fuel and air
  2. *compression stroke*, where mixture is compressed adiabatically
  3. *ignition*, where mixture explodes (ignited by spark plug)
  4. *power stroke*, where hot gases push piston out.
  5. *exhaust stroke*, where incoming piston pushes out exhaust
- C. The theoretical efficiency of such an engine is $1 - (V_b/V_a)^{\gamma-1}$
  1. V_b = minimum cylinder volume, V_a = maximum, γ as for adiabatic
  2. this efficiency is *smaller* than second law limit

IV. THE REFRIGERATORS
- A. A Carnot or Stirling engines running in reverse acts like a refrigerator
- B. Practical refrigerators take advantage of a phase change
  1. evaporator piping inside refrigerator has low internal pressure
  2. fluid vaporizes at this low pressure, absorbing heat
  3. a compressor pumps this vapor into piping outside the refrigerator
  4. the vapor condenses back to fluid there, giving up heat
- C. More practical because phase change can store very large U per kg.

V. THERMOELECTRIC DEVICES
- A. How a thermoelectric generator operates
  1. electrons have different effective potential energies in materials
  2. if the junctions at opposite ends of a material have different temperatures, the effective potential energy changes can be different
  3. leads to a net entropy gain when electrons to flow in a circuit
  4. this requires absorption of heat at the hot junction and rejection of heat at the cold junction
  5. but it leads to generation of a current from temperature difference
- B. Such generators are used where small size and reliability are important
- C. If a current is forced in the other direction, it becomes a refrigerator

GLOSSARY

Carnot cycle: an idealized sequence of ideal-gas processes that produces mechanical energy from heat at the maximum efficiency allowed by the second law.

Stirling engine: a practical and simple heat engine that uses a gas as a working substance.

internal (versus **external**) **combustion engine**: an engine where the fuel is burned *inside* the engine cylinder (like an automobile engine) as opposed to where the fuel is burned *outside* the cylinder (like the Stirling engine).

intake stroke: the part of the automobile engine cycle where fuel and air is pulled into the cylinder by the retreating piston.

compression stroke: the part of the automobile engine cycle where the fuel and air is compressed in the engine cylinder by the advancing piston.

power stroke: the part of the automobile engine cycle where the hot gases produced by the ignition of the fuel and air push the piston strongly down as they expand.

exhaust stroke: the part of the automobile engine cycle where exhaust gases are pushed out of the cylinder by the advancing piston.

two-stroke (versus **four-stroke**) **engine**: an engine where the exhaust and intake steps are combined when the piston is at its lowest point between the power and compression strokes (these steps take place during separate strokes in the four-stroke engine). Doing this allows the entire engine cycle to take place within one complete cycle of the piston (two strokes) as opposed to two complete cycles (four strokes). Two-stroke engines are often lighter (per unit power) than four-stroke engines, but are less efficient and much dirtier.

compression ratio: the ratio of the maximum volume of the engine cylinder (when the piston has been pushed out as far as it goes) to the minimum volume.

compressor: the pump in a phase-change refrigerator that circulates the working fluid and (by maintaining a pressure differential) causes the working fluid to vaporize and condense.

evaporator: the part of a phase-change refrigerator where the working fluid vaporizes, absorbing heat from the interior of the refrigerator.

condenser: the part of a phase-change refrigerator where the working fluid condenses, releasing heat outside the refrigerator.

expansion valve: the valve between the condenser and evaporator of a phase-change refrigerator that allows fluid to pass but maintains a pressure difference between the condenser and evaporator.

theromoelectric generator: a device that exploits the way that the temperature-dependent ground state for conduction electrons differs for different materials to convert heat energy directly into electrical energy.

TWO-MINUTE PROBLEMS

T9T.1 What is the main reason for studying the Carnot cycle?
A. It involves only the simplest kinds of gas processes
B. It is an example of a cycle that is maximally efficient
C. Carnot engines are commonly used
D. It serves as an archetype for gas-process engines
E. A and B
F. A, B, and D
T. none of the above (specify why, then)

T9T.2 (For discussion.) List advantages and disadvantages of the Stirling engine as a practical heat engine.

T9T.3 Even an ideal automobile engine is less efficient than an ideal Carnot engine operating between the same two extreme temperature differences (T or F).

T9T.4 A certain automobile engine has a compression ratio of 5.4. The adiabatic index γ of air is 1.4. What is the maximum efficiency of this engine?
A. 3.9   D. 0.91
B. 0.26   E. 0.49
C. −0.96   F. other (specify)

T9T.5 If we turn the flywheel of a Stirling engine by hand *opposite* to the direction it would turn if running as an engine, it becomes a refrigerator that draws heat energy out of the top end of the cylinder in Figure T9.3 at the other end. If we turn it by hand in the *same* direction that it would run as a heat engine (but without heating it) what happens?
A. nothing: the engine cancels itself out
B. it still works as a refrigerator, with the top end of the cylinder being the cold end (as before)
C. it still works as a refrigerator, but the hot and cold ends are flipped.
D. something else (specify)

T9T.6 Almost all refrigerators employ a working substance that undergoes a phase change during the refrigerator cycle. Is it possible to construct a phase-change heat engine that essentially is the refrigerator process in reverse?
A. completely impossible.
B. In theory yes, in practice, no.
C. Yes: a car engine is an example (since the gasoline goes from liquid to gas in the process)
D. Yes: a steam turbine is an example
E. both C and D

T9T.7 (For discussion.) List advantages and disadvantages of the thermoelectric generator as a practical heat engine.

HOMEWORK PROBLEMS

BASIC SKILLS

T9B.1 A certain automobile engine has a compression ratio of 6.2. What is its maximum theoretical efficiency?

T9B.2 A certain diesel engine has a compression ratio of 15. What is its maximum theoretical efficiency?

SYNTHETIC

T9S.1 Think of some applications where a Stirling engine might be useful and a gasoline engine could not be used.

T9S.2 Verify equation T9.9 as follows.
(a) Argue that heat energy enters the system in Step 3 and leaves the system in Step 5 and does not enter or leave during any other step.
(b) Argue that

$$|Q_H| = \tfrac{n}{2}Nk_B(T_d - T_c), \quad |Q_C| = \tfrac{n}{2}Nk_B(T_e - T_b) \quad (T9.12)$$

where n is the number of degrees of freedom in the gas molecules involved in the adiabatic expansions.
(c) Use this equation to show that:

$$e = 1 - \frac{T_e - T_b}{T_d - T_c} \quad (T9.13)$$

(d) Apply equation T9.5b to the adiabatic processes in the automobile cycle and show that:

$$\frac{T_e - T_b}{T_d - T_c} = \left(\frac{V_a}{V_b}\right)^{\gamma-1} \quad (T9.14)$$

(e) Use this to verify equation T9.9.

T9S.3 Show that the coefficient of performance for a Carnot refrigerator is the maximum possible allowed by the second law of thermodynamics.

T9S.4 A perfect refrigerator would move a certain amount of heat from the cold reservoir to the hot reservoir without using any mechanical energy. Argue using energy flow diagrams like the one shown in Figure T9.6b, if you had a perfect engine, you could combine it with an ordinary refrigerator to create a perfect refrigerator.

T9S.5 A perfect refrigerator would move a certain amount of heat from the cold reservoir to the hot reservoir without using any mechanical energy. Argue using energy flow diagrams like the one shown in Figure T9.6b, if you had a perfect refrigerator, you could combine it with an ordinary heat engine to create a perfect heat engine.

RICH-CONTEXT

T9R.1 Imagine that you are trying to construct a practical Carnot engine that will deliver 1 hp (700 W). The parameters are as follows. For safety reasons, your high-temperature reservoir can be no higher than 535°F (280°C). Your engine is cooled by the surrounding air. Because volume changes have to be reasonably quasistatic, you are limited to 2 cycles per second. The metal used for your cylinder can withstand 25 atm of pressure, and various considerations limit the ratio of your maximum to minimum volume to be no more than 10. Find the minimum and maximum volumes for your cylinder. (For comparison, a gasoline engine with a cylinder volume of a few hundred cm^3 would produced hundreds of horsepower.) [*Hint:* $W_{net} = e|Q_H|$, and $|Q_H|$ is given by equation T9.2c. This will get you started.]

ADVANCED

T9A.1 Consider the following heat engine. Helium (a monatomic gas) at temperature T_C is confined to a cylinder by a piston connected to a flywheel. At the instant that the piston is pushed furthest into the cylinder (so that the gas has its minimum volume V_a, the gas is given a sudden blast of heat which raises its temperature almost instantaneously to T_H (call this step 1). The gas then expands against the piston adiabatically until its temperature falls again to T_C (call this step 2). When the piston is at its farthest position (so that the cylinder has its maximum volume V_b), a valve opens and bathes the cylinder with water at temperature T_C. This keeps the gas at temperature T_C while the flywheel compresses it back to the initial volume. Find the efficiency of this engine as follows.
(a) Show that both the heat coming into the gas in step (1) and the work done by the gas in step (2) have absolute values equal to:

$$|Q_1| = |W_2| = \tfrac{3}{2}Nk_B(T_H - T_C) \quad (T9.15)$$

(b) Use equations T3.10b and T9.5b to show that the work done on the gas during step 3 has the absolute value

$$|W_3| = \tfrac{3}{2}Nk_B \ln(T_H/T_C) \quad (T9.16)$$

(c) From these results, calculate the efficiency.

ANSWERS TO EXERCISES

T9X.1 The argument is essentially the same as for equation T9.2c. During the isothermal compression of step 3, work $W_3 = -Nk_BT_C \ln(V_{final}/V_{initial}) = -Nk_BT_C \ln(V_d/V_c)$ flows into the gas (according to equation T3.10c). Using the fact that $\ln(V_c/V_d) = -\ln(V_d/V_c)$, we can get rid of the minus sign, yielding equation T9.3.
T9X.2 (Divide the first of equations T9.5c by the second and take the $\gamma{-}1$ root of both sides.)
T9X.3 Equation T9.6 means that the two logarithms in equation T9.4 have the same value and can be canceled out.
T9X.4 $(V_b/V_a)^{\gamma-1} = 8^{2/5} = 2.3$, $1 - 1/2.3 = 0.56$.
T9X.5 Equation T9.5b tells us that for an adiabatic expansion, $TV^{\gamma-1} = $ constant, meaning that

$$T_{init}V_{init}^{\gamma-1} = T_{final}V_{final}^{\gamma-1} \implies \frac{T_{final}}{T_{init}} = \left(\frac{V_{init}}{V_{final}}\right)^{\gamma-1} \quad (T9.17)$$

Applying this to step 4, we get $T_e/T_d = (V_a/V_b)^{\gamma-1}$ and to step 2 we get $T_c/T_b = (V_b/V_a)^{\gamma-1}$. Inverting the first of these and comparing with the second, we get the desired result.
T9X.6 Plugging equation T9.10 into equation T9.9 we get

$$e = 1 - \frac{1}{T_c/T_b} = 1 - \frac{T_b}{T_c} < 1 - \frac{T_b}{T_d} \quad (T9.18)$$

since $T_d > T_c$. The last quantity is the second-law limit, since T_d and T_b are the extreme temperatures for this case.

INDEX TO UNIT *T*